W9-COS-554

FLORIDA STATE
UNIVERSITY LIBRARIES

OCT 21 1998

TALLAHASSEE, FLORIDA

What Works

What Works

A Decade of Change at Champion International

Richard Ault

Richard Walton

Mark Childers

Jossey-Bass Publishers • San Francisco

HD
9829
.C5
A95
1998

Copyright © 1998 by Jossey-Bass Inc., Publishers, 350 Sansome Street, San Francisco, California 94104.

All rights reserved. No part of this publication may be reproduced, stored in a retrieval system, or transmitted, in any form or by any means, electronic, mechanical, photocopying, recording, or otherwise, without the prior written permission of the publisher.

Jossey-Bass books and products are available through most bookstores. To contact Jossey-Bass directly, call (888) 378-2537, fax to (800) 605-2665, or visit our website at www.josseybass.com.

Substantial discounts on bulk quantities of Jossey-Bass books are available to corporations, professional associations, and other organizations. For details and discount information, contact the special sales department at Jossey-Bass.

For sales outside the United States, please contact your local Simon & Schuster International Office.

Manufactured in the United States of America on Champion paper.

Library of Congress Cataloging-in-Publication Data

Ault, Richard.
  What works : a decade of change at Champion International /
Richard Ault, Richard Walton, Mark Childers. — 1st ed.
      p.   cm. — (The Jossey-Bass business & management series)
    ISBN 0-7879-4181-6 (acid-free paper)
    1. Champion International Corporation—Reorganization.   2. Paper
industry—United States—Case studies.   3. Organizational
change—United States—Case studies.   I. Walton, Richard E.
II. Childers, Mark, date.   III. Title.   IV. Series.
HD9829.C5 A95 1998
338.7'6762'0973—ddc21   98-19730

FIRST EDITION
HB Printing   10 9 8 7 6 5 4 3 2 1

The Jossey-Bass
Business & Management Series

To Pennie, Sharon, and Leslie

Contents

Foreword:
A Remarkable Journey

The Champion story of cultural transformation is compelling and timely, even though it is still unfolding. In the early 1980s the company embarked upon a determined journey of behavioral, technological, and cultural change. This book chronicles that decade-long journey through the eyes of three different participants and observers. As such, it is one of the few well-documented case histories of major change.

Over the years, bits and pieces of the Champion story have appeared in books, articles, and financial analysts' reports. Many persistently criticized the company's modest financial results and inability to counter the difficult industry trends and obstacles it faced. Others (as I have on several different occasions) have drawn on the company's experience to illustrate the power of real teams and the ability to change leaders down the line. It is certainly true that the company has left a bimodal performance trail. While its operating performance has improved consistently to set the pace for mill operations within the industry, its financial performance has lagged. As a result, many continue to overlook and downplay what the company is accomplishing in both cultural change and operating performance improvement. It is high time to correct that oversight.

This book is about broad-based change and team performance. Clearly, it is worthwhile reading for practitioners and students of change, because of what one can learn from Champion's journey. It is also a remarkable story of how an unlikely leader discovered and applied the secret of real team performance at multiple levels. Those familiar with former CEO Andrew Sigler still wonder how such a strong-willed leader could have spawned a team-based change effort that was *not* fueled by a crisis, that involved people from every level and function, and that demanded real-team behaviors in all parts of the business. Surely, this must be a writer's fantasy.

Let me assure you it is not. The old image of the company, while not without some basis in fact, hides the persistent reality of broad-based changes that are producing valuable benefits for shareholders, customers, and employees alike:

Ten of the company's original eleven mills have been completely restructured technologically, organizationally, and socially; four of the ten can claim world-class performance capability, and the others are rapidly closing the gap.

Each support group—finance, accounting, purchasing, sales, marketing, personnel, and so on—has been restructured much like the mills where five layers of management are now closer to two. Millions of dollars of profits have resulted from these changes.

For six straight years, the company has risen up the list of *Fortune's* Most Admired Companies—largely because of the reactions of its customers and its employees.

Union relations have been completely and fundamentally transformed from the traditional adversarial model to a mutual-interest model based on the "win-win" premise.

Major divestments of low-performing assets have been achieved, and more are being targeted for divestment in the near future.

Information technology improvements have increased the productivity of both mills and support service operations.

Customer-driven quality is rigorously pursued through the organization and has been recognized by virtually all customer groups served.

Despite serious downturns of the highly cyclical industry, the workforce remains intact and strongly loyal to the company in most areas.

In short, the decade-long change has produced—and continues to produce—fundamental improvement in financial and marketplace performance, as well as employee behaviors and asset portfolios.

What was once a company that could scarcely imagine achieving "acceptable returns" is becoming a dynamic enterprise that dares to aspire to be the best in the world.

The purpose of this book—and the sincere intent of its authors—is to tell the story in a way that can help others who may feel "stuck in the pack" to find the courage to try to change. It also offers a host of practical insights, approaches, and ideas that can enable others to pursue this kind of seemingly impossible, diffuse, and comprehensive change as well. To that end, the book addresses the following issues and questions:

1. How was the foundation for such fundamental change actually laid?
2. Why did the results from the change take ten years to achieve, and why is the process still going on?
3. How did past struggles serve to whet the leadership appetite for more?
4. Which elements of the change effort were most frustrating and why?
5. Which change challenges were most critical and why?
6. What did Champion learn about this kind of change that others can apply?

The change effort is far from over, and Champion leaders are by no means declaring victory early. In fact, they believe the future change challenges to be even more daunting than the ones they have been through. At the same time, their confidence, capabilities, and commitment have increased. They are battle-hardened veterans now, and they fully intend to continue "staying the course."

The uniqueness of this effort is what makes it powerful. The Champion experience is not merely instructive—it breaks new ground in several arenas. For example, there are few documented cases of major change that deal with unions and white collar employees as well. There are few broad-based behavioral change situations that were actually driven by such a strong-willed and domineering CEO; few that really married major technological investment with major social system investment; and almost none that were subjected to so much external criticism over so many years—and still persevered. There are almost no cases where a strong-willed CEO actually replaced himself with a home-grown real team—with the intention that they behave very differently. And there are almost no situations where a history of broad-based cultural change worked its way through operations, sales, marketing, purchasing research, accounting, staff services, and so on—before it reached the top-management level.

Most important, perhaps, there are few where a mediocre performer was able to change enough to have a credible shot at becoming what Collins and Porras (authors of *Built to Last*) call a "visionary leadership company"—and what Champion now calls "becoming the best." That alone would make this a one-of-a-kind story.

May 1998   JON R. KATZENBACH
Dallas, Texas

Preface

This book is about the management of a radical transformation of a mediocre company over slightly more than a decade. The transformation process integrated many of the promising approaches to organizational improvement that were current during that decade yet avoided faddishness, one-shot fixes, or commitment to any one orthodoxy. Instead, it was based on the pragmatic commitment to do "what works" within a guiding set of social values.

Champion International Corporation entered this period in 1986 with below-average physical plant, a workforce managed by traditional control and compliance techniques, conventionally adversarial union relations, a so-so reputation on Wall Street based on a record of poor financial performance relative not only to the S&P 500 but also to other pulp and paper companies.

At the end of the period covered by this book—December 1997—the picture was dramatically different. We believe Champion's decade of effort at making fundamental change is impressive, unique, and instructive. This is the story of an extraordinary transformation of a U.S. corporation into an entity that has become significantly more competitive in its industry and remarkably healthier as a human organization in spite of difficult struggles both internally and economically and in the face of considerable external criticism. "More competitive" refers to the company's cost and quality performance, its ability to attend to and adapt to changing market conditions, and its commitment to shareholder value. "Remarkably healthier as a human organization" refers to such indicators of broad-based cultural change as the personal and professional growth at all levels of the corporation, the increased dignity of workers, the fundamental shift in union-management relations, and the improved relations across hierarchical levels.

What is so impressive about the transformation? First, it results from a sustained and progressive change effort over more than a decade. One reader of this story of Champion said that in retro-spect there was "indefatigable will" at work at Champion. This inde-fatigable will may have been telegraphed by the unremarkable circumstances in which the change was launched. There was *no* perception at that time that there was a small window for change, and there was *no* employment of shock therapy. And although there were significant setbacks along the way, especially during the middle of this period, commitment and support for change objec-tives never wavered. Overall, during the decade change momen-tum gained more than it slowed. In other words, the company's success is not a story of any quick fixes; it is the result of over a decade of hard work, and, it is hoped, it has been "rebuilt to last," to borrow from the title of an important book, *Built to Last* by James Collins and Jerry Porras.[1] Collins and Porras distilled the common success factors in eighteen "visionary companies" that have sustained records of excellence since their founding. Theirs are stories of continuity. The Champion story is about instilling many of these same success factors in a corporation that had underperformed for decades; it is about achieving radical and fun-damental change. There are few (Collins and Porras would say there are no) documented cases of a mediocre performer's being able to change enough to have a credible shot at becoming a visionary company. Explaining the hows and whys of the sustained support for such change is part of what makes the story instructive inasmuch as U.S. managers are so often accused of adopting the latest management fad only to replace it later with a new one.

Second, the transformation is remarkable for its scope. Start-ing with the reform of the shop-floor work organization in one mill, the change effort broadened to include not only the other ten pulp and paper mills in the company but also its forest-products activities, its downstream distribution businesses, and its technology center and other professional units such as informa-tion systems and accounting. Over time it affected every link in the value chain—the sequence of value-adding organizational activities—from dealing with suppliers to serving customers. Remarkably, these changes occurred at the same time as a rare and impressive transformation of adversarial labor relations into a union-management partnership. Finally, and especially significant, the changes at each organizational level led to change at a higher level and reinforced the changes already started at the lower levels; by the end of the decade the organization was being transformed from bottom to top. This story stands in contrast to the typical change story, which concentrates on a particular plant or division, a particular professional activity, or a specific level of the organization.

Third, the transformation is significant because of the forces triggering it and driving it, which stand in contrast to conventional management wisdom and much social-change theory. The conventional view is that a crisis, threat, or other significant event is a necessary (if not sufficient) precondition for major change initiatives to be successful. This Champion story does not include (a) threat triggers, such as the dramatically increased global competition that has driven major change in auto plants and steel mills, (b) revisions in the rules of the game, such as the deregulation of the airline industry, which has stimulated a variety of organizational reforms in air carriers, or (c) the replacement of top managers, such as preceded major change in the publicized transformations of General Electric, IBM, and Kodak. Change at Champion occurred in an industry characterized by large cyclical ups and downs but not threatened by any sharply changed competitive forces or conditions, and it was driven for a decade by a chief executive officer, Andy Sigler, who had already served in that capacity for over ten years. As will be seen, Sigler's sustained leadership and support were absolutely critical to the success of the change effort, but the story we will tell is not the story of one man's heroic triumph, as are many books about company turnarounds. The Champion story is about engineering systemic change by recognizing and utilizing many minor threats and opportunities that could in combination support major change. As will be seen from the description of the new executive team in the Prologue, this brand of systemic change not only survived the transition from Sigler and other leaders but, building on a firm foundation, progressed to new levels.

Champion International, with 1997 sales of almost $5.74 billion, is a large player in the U.S. forest-products industry, with, in 1997, eleven pulp and paper mills, several wood-products mills, and 5.3 million acres of timberlands. In addition, it owns subsidiaries in Canada (Weldwood) and Brazil (Champion Papel e

Celulose). Headquartered in Stamford, Connecticut, with administrative offices in Hamilton, Ohio, it employs about 24,400 people worldwide and has the capacity to produce over six million tons of paper, paperboard, and market pulp a year.

From its roots in Hamilton, Ohio, in the 1890s, Champion developed and expanded in a number of ways. The focus of this book is on the process of change that began after its merger with St. Regis Corporation in 1984, from which point the business focus of the company has been on the production and sale of commodity printing, publication, and newsprint grades of paper. With the merger and the new business focus came a two-pronged improvement strategy of investing in state-of-the-art equipment and facilities and investing in the people who ran that equipment and made those facilities work—in essence, a sociotechnical change process (as described in Chapter One). This process has persisted, unlike many short-lived efforts in other organizations, for over ten years. In this book we will present the story of this decade of change and propose the key ingredients that appear to explain Champion's transformation.

If the Champion story is so impressive and instructive, why wait so long before publishing this account? Several factors explain the timing of this book. First, prior publications have documented achievements in particular parts of the organization. An article targeting managers of technical and professional employees, published in 1992, described changes in Champion's technical center—a flatter organization, self-directed teams, a technical career ladder, and the integration of technical programs with business priorities.[2] (See Chapter Two for a detailed description of these results.)

Chapters from two other books aimed at practitioners and scholars of labor-management relations focus on the Pensacola, Florida, mill management's skillful combination of "forcing" tactics to achieve substantive concessions in work roles and economic terms and "fostering" tactics to create not only a committed relationship with employees but also a cooperative relationship with union leaders. The challenges faced by the Pensacola management are typical of those faced by most U.S. managers of unionized operations, and the Pensacola case study is an example of effective handling of the dilemmas inherent in combining forcing and fostering initiatives.[3]

Another publication about the experience at Champion was authored by Jon Katzenbach. In the first chapter of his book *Teams at the Top,* he uses the transition at the senior executive level of Champion from the "single-leader working group" headed by Sigler for nearly twenty years to the "real team" created by Sigler's successors (see Prologue) to argue that creating real teams at the top is a "tougher game" than doing so at lower levels. However, he credits "the groundbreaking work with teams at the mills" with creating "a remarkable base of team experience and knowledge, which strongly influenced team efforts at the top as well as the operating performance of the mills."[4]

A second factor explaining the timing of this book has been the same long-term perspective that underlies the story itself. Management wanted to avoid premature publicity of its change initiatives.

A third factor is that the story continued to unfold, requiring us to repeatedly incorporate new chapters in the company's life. Despite the long period of time we cover, the story is and must inevitably be incomplete. As people at Champion say, "It is a race without a finish line."

Who Should Read This Book

This book is intended primarily for practitioners and secondarily for students of organizational change. Managers and union leaders who are involved at any level in major change efforts or are considering launching such efforts should glean helpful insights from the experiences shared here. This is not, however, merely a case study of one company's efforts to transform itself, impressive though that is. We have tried to relate the story of Champion's change process in a realistic way that will give organizational leaders a feel for and a sense of the ups and downs and the complexities of such an undertaking, and we have also tried to present models and to highlight lessons learned so as to help readers discover how they might apply Champion's experiences to their own attempts to lead and manage change. These lessons may be particularly relevant for those who are contemplating major change absent a major crisis but who have the vision and determination to build something lasting. This book is not for those interested in the easy fix.

Many chief executive officers inherit an organization that is not competitive or will not remain competitive without major overhaul. This book should be of particular interest to such executives. Many other books that derive the lessons from major transformation are about rapid turnarounds through restructuring. The situations are usually urgent, and the restructuring is wholly top down and painful. A prime example of this genre is *Mean Business,* by Al Dunlap, sometimes known as "Chain-Saw Al."[5] This book on Champion is about a different transformation strategy, one that revolutionizes the organization in steps over time rather than in a highly compressed period; that relies heavily on initiatives at the bottom and middle rather than on a top-down calling of the shots; that reforms the parts to see how competitive they can become and then reconfiguring the whole rather than launching the change with a restructuring of the corporation. Another distinctive feature of Champion's approach is that it mobilizes intrinsic motivation to excel by a variety of methods (such as empowerment, direct employee contacts with customers and suppliers, open-books management, and job redesign) before augmenting these motivational forces with extrinsic incentives (such as gainsharing and production bonuses). In this regard Champion's change process provides an interesting contrast with turnarounds that are launched with major shifts in compensation to align the financial stakes of employees and shareholders.

In brief, the Champion story illustrates one approach to change, one that proved effective in its circumstances. It should be considered alongside other successful approaches documented, analyzed, and generalized by others, such as Noel Tichy's treatment of General Electric;[6] Robert H. Miles's studies of National Semiconductor[7] and of Southern Company, Norrell Corporation, and the PGA Tour;[8] and David Nadler and David Kearn on Xerox.[9] Although we believe the Champion story has broad implications for executives who find themselves in circumstances not unlike Champion's, we do not wish to argue that one size fits all. Indeed, the avoidance of any such orthodoxy is a key theme of the Champion story.

Practitioners in a position to assist change leaders should also find the book of benefit. External and internal organizational-development professionals, human resource managers, and others who find themselves in roles supporting major transformation efforts should find some empowering concepts, process insights, and references to tools to help them help their clients.

Finally, although this is not primarily an academic book, it should be of interest to those interested in the theory of large systems change. It should certainly be an important addition to the database in this field, and teachers of courses in management and the management of change may want to use this story to complement their other coursework, perhaps using it in conjunction with accounts of contrasting approaches to systemwide change.

The Authors: A Unique Collaboration

We believe that this book is also unique because of the combined perspectives of the three authors, one (Ault) the lead external consultant for the process, another (Childers) an executive who has been instrumental in leading the change, and the third (Walton) a professor in the field of strategic change at the Harvard Business School and also a member of the Champion board of directors through most of the period covered by this book.

Thus, all three of us are participant observers—albeit with different participant roles and different angles of observation—and our product can be thought of as "action research," to use the parlance of the social sciences. (For a description of our approach to this project, see Appendix A.) We set out on this joint project because, although the Champion process was not without tactical failures and missed opportunities, it was on the whole, we believe, extraordinarily successful. Two of the authors have been associated as consultants with many change efforts—of varying degrees of effectiveness—and they single out this experience as the best example of putting together all the pieces of effective companywide change efforts. It is not that the accomplishments of any of the individual units are novel (although the success in the technology center, the process at the top management level, and the developments in the union-management relationship are especially noteworthy) or that any of the change processes broke new ground conceptually or in technique. Rather, the overall success of the project can be explained by the way these processes were timed, sequenced, executed, and combined.

How This Book Is Organized

The Prologue describes developments that occurred at the top of the organization toward the end of the roughly ten-year period covered here—a new chief executive officer and other members of his "Gang of Eight" take the change effort into new territory. This transition in 1996–1997 could only have occurred because of the changes that preceded it. The rest of the book describes these changes.

In Chapter One we provide a brief overview of the Champion change effort, identifying key developments, assessing the amount of organizational change—including change in mind-sets, behavior, and relationships—and analyzing the performance results that can be attributed to these changes. Then we elaborate on two crucial themes of the story, themes that set the Champion experience apart: the comprehensive nature of the change and the sustained nature of the change effort. We conclude with a summary of actionable ideas the reader can take away from Chapter One and with an explanation of the conceptual framework we use in the remainder of the book.

The description in Chapter One of a sustained, comprehensive, and highly successful change effort begs two questions: How did the company manage the change process? And why was the change so successful? These are the questions the next three chapters address. The roles chapters Two, Three, and Four play in this story are derived from a simple conceptual framework that identifies what we regard as the key ingredients of successful change at Champion. We believe this framework could also prove useful to many other organizations facing the need for change in today's turbulent business environment. We developed this framework largely from interviews conducted at all Champion's U.S. paper mills and at selected other locations between 1991 and 1993, approximately five to seven years into the change process. By this time considerable experience with planned change had accumulated in the organization, and this assessment was an attempt to learn what worked and why it was working. In group interviews at each hierarchical level at these locations, people were asked to reflect on all the different strategies and approaches encompassed in the change process and to think of a time things "really clicked," of a time when they went home at the end of a shift or a week or a project and thought, "That's the way it ought to work all the time." After sharing the anecdotes, they were asked to identify the underlying factors behind these successes, the conditions that made them work. A pattern emerged that ran through these stories from every level and every location.

This pattern encompasses *aligning* mind-sets and behavior consistent with change objectives (Chapter Two), building the individual and organizational *capability* required for the newly envisioned corporation (Chapter Three), and *letting go* of supervisory practices, functions, and other activities that could be assumed by subordinate individuals and units (Chapter Four). This consistent pattern suggested that although there was no one right way, there might indeed be a common framework that encompassed all of the right ways.

Having begun with a prologue of scenes from the recent past, we conclude the book with an epilogue describing the direction in which Champion appears to be headed in the future.

Acknowledgments

Many change efforts fail early on because senior executives don't support or get involved in the process or they seriously underestimate the resources—including time, both in short-term intensity and long-term commitment and patience—needed to get the job done. In our case, Andy Sigler and Whitey Heist (CEO and COO respectively) provided exceptionally intrepid change leadership for over a decade. While they assiduously avoided publicity for these efforts over those years, we were able to persuade them of the importance of sharing Champion's experience with others through this book. A second point of particular vulnerability in change efforts occurs when supportive top level leadership leaves for one reason or another and is replaced by leaders who, either by omission or commission, allow or cause the decline of the process. Again, Champion is a notable exception. When Sigler and Heist retired in 1996, Dick Olson and Ken Nichols, two longtime change advocates in the company, assumed the top positions and not only supported the ongoing change work but sparked an acceleration in the change curve. They too have supported and been helpful in the preparation of this book. Together with their executive teams, Sigler, Heist, Olson, and Nichols have provided the kind of change leadership that change agents everywhere covet and seldom find.

But support did not come only from the top. Indeed, when we think of those to whom we are indebted in the making of this book, we think first of those "Champions" at every level, in every kind of job—executives, managers, workers, union officials—who worked and struggled and made change happen over the ten years and more encompassed in this remarkable story. Many of them shared their experiences and insights with us as we developed the book. Without them, of course, there would be no story to tell, and to them we extend our heartfelt thanks.

We also wish to thank our many colleagues who encouraged us, including Jim Taylor of the University of Southern California, who first suggested that we write this book. Among those who have read and commented on parts or all of the manuscript, we especially wish to thank Rich Cherry, Pam Posey, and Jon Katzenbach as well as our editor at Jossey-Bass, Cedric Crocker. The final result is, we think, significantly improved because of their feedback. We also appreciate the Work in America Institute for encouraging and publishing sections of this book as part of their PAR (Participation, Achievements, Rewards) research project.

Finally we thank Margo McCool and Margaurite Dole at the Harvard Business School for their work in preparing early drafts and Susan Lionetti of Champion for her coordination efforts in the typing and putting together of the final manuscript.

May 1998   RICHARD AULT
                                                                  Navarre, Florida

                                                                  RICHARD WALTON
                                                          East Orleans, Massachusetts

                                                                  MARK CHILDERS
                                                            Stamford, Connecticut

The Authors

RICHARD AULT has consulted for over twenty-nine years in the field of organization change. From strategic planning and team building at the executive level to work redesign in plants and offices, his work in the systematic transformation of complex organizations has taken him into both the public and private sectors in the United States, Canada, Mexico, Australia, Singapore, and Malaysia. Director of North American Quality of Work Life Activities for General Motors Corporation in the 1970s, Ault has consulted privately since 1980 with such companies as Alcoa, Appleton Paper, Bethlehem Steel, Clark Equipment, Champion International, James River, Union Carbide, Volkswagen, and such public sector clients as hospitals, universities, school systems, the State Electric Commission of Victoria (Australia), the Sydney Water Board (Australia), and the Executive Office of the President of the United States. Ault was the lead consultant in the Champion change process covered in this book.

Before entering private industry as a consultant, he served in public education as a teacher and principal at the secondary level, and taught at Western Michigan and Wayne State Universities. He was also adjunct senior consultant for the American Productivity and Quality Center in Houston. He received his B.A. and M.A. degrees from Central Michigan University and his Ed.D. in Educational Leadership from Western Michigan University.

A member of the Organization Development Network, the STS Roundtable, and the American Society of Training and Development, he was the first recipient of the annual Practitioner of the Year Award given by ASTD's Quality of Work Life Committee in 1988.

RICHARD WALTON is the Wallace Brett Donhan Professor Emeritus at Harvard University Graduate School of Business Administration. His recent interests have included social innovations that elicit high employee commitment, enhance business performance, and promote human development. He has been active as an architect of these innovations and as a researcher.

Walton earned his B.S. and M.S. degrees at Purdue University and his doctorate at the Harvard Business School. He recently retired from the faculty of the Harvard Business School after twenty-nine years of service. He has consulted with many leading American corporations and served as director of several corporations. He is currently a director of Champion International Corporation.

He is the author of a dozen books and over eighty articles. One recent book, *Innovating to Compete: Lessons for Diffusing and Managing Change in the Workplace* (Jossey-Bass), compares and explains the workplace innovation records of seven countries including the United States, Japan, and five European maritime countries. Another book, *Up and Running: Integrating Information Technology and the Organization* (Harvard Business School Press), presents a practical theory for effective implementation of advanced information technology. One of Walton's latest coauthored books, *Strategic Negotiations: A Theory of Change in Labor-Management Relations* (Harvard Business School Press), proposes a theory of strategic negotiations illustrated by contemporary developments in labor-management relations intended to increase the competitiveness of American firms while also promoting the interests of labor. This book includes an analysis of labor-managment developments in the pulp and paper industry and presents a case study of one of Champion's mills.

MARK CHILDERS is senior vice president-organizational development and human resources of Champion International Corporation. In this position, he is responsible for the company's human resource programs and its extensive organizational change efforts throughout the company.

Childers joined the company in 1978 as a senior process engineer at the company's newsprint mill in Lufkin, Texas. Since then, he has held a variety of technical and management positions in the company. In 1991, he was named corporate vice president of organizational development projects, and was named senior vice president in 1992.

A native of Dallas, Texas, Mr. Childers earned a B.S. degree in chemistry and mathematics from Stephen F. Austin State University.

Prologue

*Without the previous ten years of change, we would not
have been able to conceive of the notion that teams could
run the business. Our experience with teams and the change
process as a whole allowed us to think out of the box.*
—Dick Olson, Chief Executive Officer, August 1997

During the summer of 1997 if you were that celebrated fly on the
wall and found yourself in the boardroom on the top floor of
Champion International Corporation headquarters in Stamford,
Connecticut, you might have observed a dynamic group at work.
Known as the Gang of Eight, or, more officially but less colorfully,
as the executive committee, the members of this group had been
meeting frequently that whole year to reexamine their corporate
strategy, and they were now coming toward the end of that process.
They faced a self-imposed deadline of early October to announce
their plans to Wall Street and to their own organization. The
announcements were expected to be important and were eagerly
awaited both internally and externally. Perhaps it doesn't seem
unusual for corporate strategy to be discussed in a corporate
boardroom, but such discussions are not quite as common as one
might think or as they probably should be.

However, more unusual than the content of the agenda is the
way the group is working together. The phrase *new executive team* is
usually nothing more than an empty cliché, meaning only that new
individuals have been appointed to the top of a set of functional
silos in a bureaucratic, hierarchical management structure. It is
true that this top-management team has been in office slightly less
than a year (although each member had long service with the company before being appointed to his present job), but what you are
observing in the Champion boardroom is a management team that
is actually working as a team. And that is rare.[1]

1

Creating a Team

The Gang of Eight replaced the previous, more bureaucratic executive committee of twenty-two managers of loosely knit functions. Dick Olson, named the new chief executive officer (CEO) in the summer of 1996, had long been a people-oriented manager who stressed teamwork in all his many previous roles in the company and who in 1986 was one of the authors of a study on the operation of the Champion paper mills that recommended a participative, cross-functional, and team-oriented approach to the management of those mills. In 1985 he was instrumental in the planning and start-up of Champion's first team-based, high-performance paper mill, and soon after, in early 1986, he helped initiate the first retrofitting of an existing mill for high-performance teams. He led the redesign of all the technical staff groups into an interfunctional, team-based applied-technology organization in his most recent assignment as executive vice president for these staffs.

Named at the same time as Olson to what was essentially the "office of the chief executive" was his partner, vice chairman Ken Nichols, who also brought to the table substantial credentials as a leader in the movement over the last decade toward a team-oriented culture at Champion. He was the sponsor in 1987 of the first high-performance structure in a support-staff area, the accounting function, and of a major change in the corporate technology center at about the same time. This project in the technology center, discussed briefly in the Preface, resulted in a substantial flattening of the structure, with scientists working in self-managed, cross-disciplinary, business-oriented teams formed around their internal customers' needs. It was a precursor to and prototype for the larger-scale redesign of applied technology later led by Olson. Nichols had encouraged change along similar lines in many corporate staff groups at the Knightsbridge administrative center in Hamilton, Ohio, and he employed a human resources development system among the staff groups reporting to him that was the envy of others in the company, including the human resources department itself. Both Olson and Nichols were instrumental in seeing that Champion's change process of the previous decade reached comprehensively across the entire organization rather than focusing totally on the paper mills.

Even though they didn't officially take office until October 1996, they brought the Gang of Eight together over the preceding summer to look at the transition that would be brought about by the retirements of Andy Sigler, who had been CEO for over twenty years, and of President and Chief Operating Officer Whitey Heist. Another five of the top twenty-two managers were due to retire over the next few months. The group met almost daily with Sigler's encouragement but—significantly and to his credit—without his strong presence, even though, as Nichols said, "it damn near killed Andy not to be there." This newly formed transition management group at that time included the heads of Champion's three major paper business units, its forest-products business unit, and the functions of sales and of organizational development and human resources, plus Olson and Nichols. Some of these men had been leading candidates for the jobs to which Olson and Nichols were named, and Nichols observed that "the first few meetings were awkward, with some of us still nursing our wounds. But we eventually got to having real good discussions." Olson, whose basic belief is that "none of us is as smart as all of us," expressed amazement at "how quickly we came together as a team." Mark Childers, senior vice president of organizational development and human resources and one of the authors of this book, commented, "We went through the stages of team development pretty quickly. I never saw a group come together quite as well."

Olson's style figured prominently in this early gelling. Joe Donald, executive vice president of the successful publication papers business unit when this process started, observed that "Dick is probably the most open of all of us." Open himself, he encouraged openness in others and, along with it, an open-mindedness in listening to others, in reflecting on decisions and practices of the past, in considering hard information as opposed to past prejudices, and in weighing options for the future. He also encouraged a sharing of leadership depending on the topic, setting the tone for a dynamic process for which he would set the agenda and provide closure but which, in between, made for free-flowing dialogue. His leadership style according to Mike Corey, senior vice president of corporate analysis, was to "set expectations, encourage free thinking, and then, as we near closure, to encourage consensus. If we can't reach consensus, he will decide."

In addition to Olson's style, another factor that contributed to creating a real team was that the members of the Gang of Eight had for the most part grown up together in Champion as colleagues. This collegial relationship was in direct contrast to their experience under Sigler, who had many more years of experience than they did, a fact that contributed to or reinforced (or both) his more naturally autocratic style. Twig MacArthur, then executive vice president of newsprint and kraft papers, suggested that "Dick had the authority to come in and become a dictator, but he didn't. All of us, including Dick himself, are aware of what we don't know. We grew in the business together; there is no one like Andy who can say he knows a lot more than the rest of us or is older and wiser." MacArthur believed that they now interacted more like brothers, adult to adult, whereas before it had been more of a parent-child environment.

Creating a More Horizontal Structure

That transitional summer they came to agreement on priorities for the future: to focus on operating excellence and on growing the business and to continue to focus on people, which had been at the forefront for the previous decade. They would continue to support the principles that provided the foundation for their culture as embodied in a statement called "The Champion Way." Having committed to these basic priorities, they began to wrestle with structure: Will the existing structure support these priorities, or do we need to change it? They had been part of a functional structure back in the mid-1980s, with highly centralized manufacturing and sales functions. At the time that structure was found to have serious flaws, and the operations study team of which Olson was a part recommended to Sigler in 1986 that the functional structure be replaced by interfunctional business units, with a strong emphasis on moving profit responsibility to a lower level. Sigler implemented their recommendations within a month; the business units had been the core of the operations ever since and were seen as a major improvement.

Key members of the transition group were the business-unit executives, and they were strong supporters of that concept. Olson himself had been one of the original business-unit heads, and even though he had been working in a staff capacity for several years since then, he still counted himself as a supporter of the business-unit structure. Nevertheless the team, in trying to take a fresh look at everything, questioned that structure as well. They recognized that with all its benefits there were also weaknesses. For example, there was a virtual absence of marketing, and business-unit boundaries too often got in the way of the potential synergy of products and customers. Through dialogue, they, as Olson put it, "came to believe that operating excellence could be better achieved if we put all the mills together. We could gain greater leverage for manufacturing and sales." As they reallocated roles and functions, a new marketing function emerged in the middle as a catalyst and integrator for sales and manufacturing. But if they centralized these functions, how would they avoid the pitfalls of the functional structure they had thrown out years ago? How would they retain some of the benefits of the business-unit structure?

The answer lay in the cultural change that had transpired in the ten years since that old structure had been in place. Over that time, people at all levels had learned to work in groups spanning structural boundaries. The interpersonal, collaborative skills required to make these groups work had become part of the competency matrix of the whole company. Team structures and processes had become institutionalized to the point of being standard operating procedure. And people, including the people in that boardroom, had learned the power of collaboration; they knew that teams weren't just a management fad but that they could be made to work. None of those conditions existed in 1986. Now these core competencies freed the group to look at overlaying the three major vertical structures—manufacturing, sales, and marketing—with a horizontal structure, replacing the three paper business units with eight "business teams," each staffed with a middle-level executive from the three areas, with the new marketing function acting in a coordinating capacity. Each business team would be responsible for strategy and profit performance in its product segment. These teams enabled more specific product, market, and customer focus than did the three previous business units, which had been designed around manufacturing technologies. For even more specific focus, the new business teams would in turn sponsor cross-functional product teams.

Uniquely, these business teams report *as a team* to what Olson refers to as the "chief operating officer (COO) team" made up of the three newly appointed executive vice presidents of the three functions. (Olson resists calling them functions. To emphasize the point that this is a different concept than in the past, he refers to them as "the vertical teams.") Ten years earlier no one would have even suggested that a team report as a group to another team. Joe Donald moved from executive vice president of publication papers to head up manufacturing; Scott Barnard had been head of sales for publication and printing and writing papers and now became executive vice president for all of sales; Twig MacArthur, executive vice president of the old newsprint and kraft business unit, took over as the leader of the new integrative marketing function.

The Gang of Eight had sufficient understanding of team processes to realize that merely announcing a new structure would not make the business teams work. They added systemic support by changing the reward system so that 40 percent of the bonuses of business-team members were now determined by how well they worked together to get the teams off and running. Further, the offices in the headquarters building were regrouped by businesses rather than by the three vertical functions. One hundred and seventy-nine office moves were made to avoid functional silos.

After the Gang of Eight made the necessary staffing decisions, the new COO team, with support from Olson and Nichols, helped get the new business teams launched in late September with a day-long workshop aimed at clarifying goals, roles, and ground rules. With help from internal consultants from the organizational-development group, these teams went on to further define their mission, roles, and processes, and got down to work. After several months, as might be expected, some had gelled successfully while others had not. Adjustments were made to the number of teams and to team membership so that almost a year later Joe Donald could say that although there was still a range in performance, the majority of the business teams were performing at the higher end. Still, said Nichols, "We put these people on the job without adequate training in analytical and strategic work. We just don't have enough people yet with those kinds of skills."

Thus a completely new corporate structure was in place and beginning to operate by the end of 1996. The business teams, as well as other issue-focused, cross-functional, cross-location study teams, also influenced upward, adding to the forces for change at the top. "They questioned everything we had done in the past," said MacArthur. "Our senior management team had to declare a general amnesty amongst us because each of us owned some of those past decisions. We could each say, '*Mea culpa,* but don't get on your high horse because yours is coming.'" Thus both internal and external pressures led the Gang of Eight to new insights about their company and their business.

Creating an Externally Oriented Mind-Set

Looking back to those early days at the helm, Olson observed, "It is amazing how little we knew about the business. To some extent Andy sheltered us all from the outside world. We were internally focused; we seemed to think we could make what we wanted, that we had good quality and service, and good facilities. What we discovered is that the external world has changed a helluva lot. My greatest learns have come through interfacing with the outside world." At the beginning of 1997 he and others on the team began to focus on long-term total shareholder return (stock price appreciation plus dividend yield), communicating that new emphasis to Wall Street (where it was immediately well received) and internally to employees. While continuing to emphasize Champion's social values, Corey said, "the single biggest cultural difference we have to have—and this is increasing every month—is more thinking as a business, as business managers, how you build a strong business over time. It used to be how you build a strong function or strong locations over time. It used to be whatever the share price was, it was. Now it's how does managing the business influence value to the shareholder. But we're still talking value over the long term not just the next quarter."

This new commitment was strongly communicated through what Nichols termed "the most basic change in compensation in a very long time." Incentive compensation for managers shifted from emphasis on earnings per share to return on capital employed, with longer-term executive compensation linked to the goal of total shareholder return. Key managers now would be required to attain designated levels of common-stock ownership, and a major part of the compensation of the board of directors was changed from cash to company stock.

At the same time, the senior team began the process of reexamining company assets and strategy within the context of this new emphasis on the external world of shareholders, customers, and competitors. Putting their new team dynamics to the test, they engaged in months of what Nichols called "the most intense, rigorous, analytical and objective piece of work that's ever been done around here." Corey describes the team as "dedicated to looking at things with a new set of eyes—more objective, without our old prejudices; more open to facts and outside advice. The single biggest change, though, is our no-holds-barred exchanges."

Such exchanges, which you may observe from your spot on the boardroom wall, are not always comfortable. "There is some tension," MacArthur observed. "We are sometimes more acrimonious; there is more questioning from peers; there are more scrimmages; a more active chemistry." Nichols commented, "We listen to each other more, but the honeymoon period has worn off, and there is more tension than I would care for in the last few months. Still there is more willingness to consider fact-based analysis and less promoting one's own agenda. We are coming [along] pretty well, but this is a tough business." MacArthur again: "There is more give and take. Nobody's feelings are spared and it's not always pleasant, but we learn a lot more. I have never known and understood our business better, and I have never been so confident in a successful future for Champion."

As the October 1997 deadline for announcing their new strategy approached, Olson reflected on his "evolving vision." "When I first got this job, I said we wanted three things: operating excellence, a continuing focus on people, and to grow the business. I still want those same things, but our understanding of what we mean by each of them has expanded and grown more complex. Grow the business has evolved to reduce cyclical earnings volatility, grow revenues profitability, earn the cost of capital over the business cycle, generate free cash flow. We will consider partnerships and alliances to accomplish these things. We have to change even more from internally driven to more customer driven. We started talking about being customer driven a few years ago, but we are a long way from where we have to be. We need to spend more time with our customers at every level, to really understand what their business is and really let that drive us. We don't pay enough attention to our competitors. Operating excellence cannot be just focused on the mills anymore; we need operating excellence in our legal staff and our tax department and all across the company. People are still the bedrock on which we will build; as what we need to do, business wide, becomes clearer, we will communicate these things to our people and give them the tools to help them do the job."

The Past Is Prologue

As a fly on the wall, then, you observe a new management group operating at an extraordinary level of "teamness" themselves, having also created a team-based management structure at the levels just below them. Several driving forces have carried them to that point. First is the natural style of the CEO, Dick Olson. Second, the members of the team, including Olson, had come up together as colleagues and were able to interact as adult peers rather than on a paternalistic basis. Third, they were faced with changing and formidable outside business conditions, which forced them for the first time in more than ten years seriously to examine the alignment of the company's technical assets, culture, and competitive business strategy. Because of the first two forces, they had the capabilities to take this task on as a real team.

In addition to these three forces, there is an even more important factor at work, an "enabling condition": the capabilities—including the competencies, the commitment, and the spirit of cooperation—that had been developed at Champion through an organizational change process that had evolved over the previous decade. MacArthur said, "Our redesign of the top management and of the business teams [was] based on our experiences with redesigns at lower levels, our experiences with cross-functional teams, and even our experience with our previous top-level 'working group.'" Perhaps Nichols put it most strongly and clearly when he said, "Without the previous ten years, we wouldn't have stood a snowball's chance in hell. It would have been a disaster."

This book is an attempt not only to tell the story of those previous ten plus years but to clarify, to explain, and to interpret what was learned from them by Champion that can be learned by others who take on the challenge of transforming their organizations for the twenty-first century. As an observer, you may want to follow that story as it brings you back up to this point, when critical decisions for the future are about to be made.

Chapter One

Dogged Persistence and Unjustified Optimism
An Overview

Are we dreaming when we talk about these things? I think not, and for this reason: We started up a new pulp mill in Quinnesec last December. And we started that mill with a nontraditional organization. . . . The whole mill runs on this team concept. . . . If we could possibly wave the magic wand and wake up tomorrow with all of our older mills retrofitted to this team concept, we'd love to do it. Our goal is to retrofit all the existing facilities to this way of doing things.
—Former President Robert Longbine in a 1986 presentation to security analysts

In an organization that has been around over a hundred years, it is difficult to declare a specific date for the beginning of a change process. After all, it had been changed and changing all along or it would not have survived. Although we date the story that is our subject from the St. Regis–Champion merger and other developments in the mid-1980s, it clearly had some of its roots in antecedent conditions dating back at least to the beginning of that decade. In 1980, Champion took a strike at its flagship mill in Courtland, Alabama, but continued to run it with salaried employees (see Chapter Three for details). From that trying experience, CEO Andy Sigler determined to pursue a better way even if he didn't know what it was. He drafted a statement of company values, "The Champion Way," that continues to be alive and well and shaping behavior to this day (see Appendix B). Executives traveled to the field locations on "Champion Way Visits" to develop understanding of these values and to encourage initiatives in support of them.

A program called Employee Involvement was already in place; it resembled the quality-circle activities widely instituted in many companies in the 1980s in which employees voluntarily engaged in problem-solving groups addressing issues in their work areas. As was the general experience with this approach in U.S. business and industry, Champion tasted enough success for it to learn that employees could make a contribution with their minds as well as their hands and backs, that teams of employees working together could indeed solve meaningful problems. However, Champion also learned the limitations of an approach that treated teams as part-time "parallel organizations" existing alongside a full-time organization that was otherwise quite traditional.

A New Strategy Takes Shape

Building on these experiences after the merger, Champion's new two-pronged sociotechnical strategy (see box) first took on tangible, physical form in a small village in Michigan's upper peninsula when the Quinnesec pulp mill started up in December 1985. Strictly speaking, this mill had been on the drawing board for a long time before the new strategy was developed and, thus, cannot be considered a product of the strategy but more a tangible, reinforcing forerunner of it. But reinforce it did, as we shall describe.

In the fall of that same year, even as Quinnesec was in final preparations for start-up, it was determined that for the strategy to work and have credibility across the organization, it had to be demonstrated in an established mill, one where it was necessary to "retrofit" the technology and the social system at the same time. For a number of reasons, the integrated pulp and paper mill in Pensacola, Florida, was chosen as the pilot. After initial successes in a *greenfield* (or new) mill and an established unionized mill, a corporatewide diffusion began in earnest. (See box for the time line of events in this sustained change process.)

The sociotechnical-systems approach was pioneered by Eric Trist and Fred Emery, beginning with their work at the Tavistock Institute in London in the early 1950s. The theory and practice are grounded in the fundamental concept that work gets done more effectively when the separate but highly interdependent social and technical systems that produce the flow of that work are "jointly optimized" rather than the more traditional "maximizing" of the technical system at the expense of the social. Among other results, this sociotechnical movement has led to the widespread use of self-managed work teams and other elements of what are now commonly known as "high-commitment" or "high-performance" organizations.

Quinnesec: Starting Up a Greenfield Mill

This new mill was conceived as early as 1979, while Dick Olson, who would become Quinnesec project manager, was the operations manager at a mill in Pasadena, Texas (since sold). Al Prendergast had been brought in as human resource development manager for the Champion Papers Division in 1977; he had benchmarked the high-performance designs of mills at Proctor & Gamble and Mead. President Robert Longbine appointed Olson project manager, and he and Prendergast conferred on the concept of a greenfield mill. A management task force formed in 1980 to consider the design of Quinnesec determined that the mill would be state of the art both in equipment and in its organization. A philosophy statement was developed emphasizing teamwork, involvement, and training.

To some extent the decision to go this way, even before Champion had determined this strategic direction for the company as a whole, was influenced by the Champion Way statement. Steve Goerner, a member of the task force, who later became the first Quinnesec mill manager, said, "We derived Quinnesec's philosophy statement from the Champion Way [statement]. It established our particular organizational values and expectations of performance—the way we should conduct our business as well as a guide for consistent policy application."

Antecedents

1974 Employee Involvement program

1980 Courtland strike

1981 Champion Way statement

1984 Merger of Champion and St. Regis

1985 Quinnesec start-up (December)

Decade of Change

1986 Major modernization capital program begins (amounting to $3.5 billion over ten years)

Pensacola product and machine conversion

Champion Way in Action statement

Business-unit structure implemented

Transition team formed

Bargaining for flexible contract language

1987–1988 Flattening of the organization

Diffusion of change process to other locations begins (includes staff units as well as mills and forestry operations)

1989–1990 Protracted and antagonistic labor negotiations at five mills

1991 UPIU and Champion sign Joint Statement of Principles

1992 UPIU–Champion Forum launched

1993 "Treetops" executive-team development process begins

Best practices and other cross-functional team processes

1991–1994 Paper-industry recession produces dramatic decline in prices (ends mid-1994)

"Steep-slope" improvements in operating performance

1995 Prices turn up sharply

Something Extra bonuses

Record earnings

Champion named "Company of the Year" in paper industry

1996 "Becoming the best" dialogue

Prices slide again

Sigler, Heist retire; new executive team named

1997 Gang of Eight team process

New functional structure and business teams implemented

New business strategy developed

The Quinnesec design task force studied innovative greenfield start-ups both outside the industry (Shell Chemical in Sarnia, Ontario; Proctor & Gamble in Oglethorpe, Georgia) and inside it (Mead in Stevenson, Alabama; Southeast Paper in Dublin, Georgia; Proctor & Gamble in Mahoopany, Pennsylvania, and Albany, Georgia) and determined that building a full-time total team culture was the only way to match the state-of-the-art facilities that were planned. However, Chairman Sigler said that he never would have approved the proposal if it hadn't been for his experience meeting with and listening to presentations by the Employee Involvement teams that had been working in the existing mills. They demonstrated to him at least some of the potential power of teams of workers.

Because of the nationwide recession of the early eighties, the Quinnesec project had to be put on hold a couple times. As a result, in December 1985 it was the first Champion mill to manifest the new strategy and to have the characteristics of what is generally thought of as a high-performance, high-commitment organization—characteristics such as self-managed work teams, pay for knowledge, and a relatively flat and lean management structure.

Employees were quick to respond favorably. As one long-time paperworker said in the first year of operation, "I worked eighteen years in other mills. I had absolutely no say in the operation. If I came up with a suggestion, they'd say, 'No, no.' Then I'd come back after a weekend off and find my supervisor taking credit for my idea. But if you made a mistake, they were real quick to tell you about it. Is this a better place to work? God, yes."[1]

At the same time, results reinforced company hopes for this new model organization. After only three months, the mill was producing more than its planned capacity, and quality was exceeding expectations. Champion's president and COO at the time, Robert Longbine, said, "We set what we thought was a short learning curve at the Quinnesec mill from the day we started up to the day we achieved rated production. It was a number of months. Actually, that mill got up to rated production within days because everybody out there was really devoted to making the full concept work and being part of a team. It was an excellent experience for us, and opened our eyes to the potential of the team system."

The experience did not turn out to be a flash-in-the-pan result of "Hawthorne effect" novelty either. It has continued to be successful over the years since, through good times and bad, through management changes, downsizing, and other challenges. The biggest such challenge was undoubtedly the start-up in 1992 of the new paper machine, making this mill a fully integrated pulp and paper mill and doubling its workforce from approximately three hundred to six hundred employees. Although there was tension because of both the new technology and the assimilation of new people into the Quinnesec culture, the mill to this day produces the best-quality product in its market, and its philosophy lives on. This sort of experience is not all that unusual with plants that operate this way. It is commonly feared that cultures of this type are fragile and that the slightest setback will cause such organizations to crash on the rocks of despair or at least to revert to a traditional mode. The actual experience is that they usually are robust, successfully weathering challenges that are thrown their way at least as well as and probably better than traditional organizations.

As Longbine said, however, it was not enough to have a Quinnesec, a greenfield success. Although it "opened . . . eyes to the potential of the team system," the goal became to "retrofit all the existing facilities to this way of doing things." Such a goal is not always achieved. Many firms have successes with greenfield plants but experience great difficulties and failures when they attempt to spread the new work systems to their established facilities. Thus, Champion management recognized that there are no magic wands, but even as Quinnesec was still in the earliest stages of start-up, the attempt to retrofit began at one of the ten unionized mills, the Pensacola mill.

Pensacola: Retrofitting an Established Mill

The Pensacola mill became a part of Champion as a result of the St. Regis merger. St. Regis had started up a major new high-volume paper machine (P5) in 1981 to make brown paper for paper bags and other uses. Technically it had been a great success; however, the brown-paper business was in a slump that threatened to be permanent, and, in any case, it did not fit Champion's new strategic profile for being in the white-paper business. Thus the new owners shut down and sold the bag-plant operations, and plans were set in motion for a major capital rebuild to convert P5 to a white-paper machine. However, before the money was totally committed, the company, as a part of regular negotiations for the 1985 union contract, demanded changes in work rules that would allow, for example, increased manning flexibility. With employee uncertainty at a high level, the company had the upper hand and was successful in "hard bargaining" the changes. (Similar agreements were negotiated at the other mills in 1986.) As one manager said, "They thought we rammed it down their throats, and we did." This was not an unusual approach at Champion or in the industry at the time; it was more or less the mode of the day. Soon after, though, came the decision to spread the Quinnesec approach, a difficult task requiring management to elicit employee commitment rather than settle for mere compliance.

Pensacola became the primary pioneer for retrofitting because it offered several advantages:

- The product conversion aligned with Champion's new business strategy would require major changes in technology and would also result in significant job changes and create opportunities to plan new procedures for staffing new equipment and new units.
- Champion had already negotiated work-rule changes to permit increased flexibility in managing operations; a participative approach offered a means for realizing the flexibility now permitted in the union contracts.
- Given the greater sense of uncertainty at this mill than at the others, labor was more easily persuaded that it was in its best interest to cooperate and make the mill more productive.
- It was a southern mill without a tradition of militant unionism; fairly congenial relations had prevailed prior to the 1985 negotiations.

In total, then, at Pensacola there was a new company with a new philosophy, a new business strategy, a new product, new technology, and a new contract. A great window of opportunity was opened for change, and several activities were started in early 1986 to promote both improved relationships and the redesign of work structures:

- Education and dialogue were initiated within mill management and between management and labor about new ways of working.
- Pilot efforts resulted in redesigning three departments to use the sociotechnical model to group jobs into work teams and to change lines of progression and wages to encourage flexibility, teamwork, competence, commitment, and self-supervision.
- In 1986 a new corporate planning mechanism called *targeting* was introduced as a central means of giving a business focus to the involvement activities. This process includes developing targeted improvements for key cost and quality drivers over which each unit has control, formulating plans to effect the improvements, and then reviewing progress toward those targets. Pensacola took this mechanism, which could have been just another "number-crunching" exercise, and made it a vehicle for participative planning and performance management, including creation of a mill mission statement and strategic business model along with detailed five-year mill and functional plans.
- Stakeholder committees of union and management officials were formed to provide oversight for those activities involving labor and management.
- Generous amounts of training and outside consultation were employed to ensure the success of these activities.
- The construction of the new facilities for the product conversion along with extensive efforts to clean up and fix up the old facilities provided continuing evidence to employees of a promising future for the mill.

These activities were only the highlights of the earliest stages of a long-term change effort for this mill (as well as for Champion as a whole), and hard results did not come immediately; however, by the summer of 1986, with the new business strategy formulated and visible early successes and feedback from both Quinnesec and

Pensacola, Champion was ready to spread the emerging organizational model to other locations and groups.

Learning What Works: Diffusion Across the Organization

The business strategy no longer existed only on paper or in managers' heads. They could go to Quinnesec and Pensacola and see it working in practice. In fact, after a tour of the Pensacola mill by some eighty senior managers not too long after the big, newly converted P5 machine was up and running, producing big rolls of commodity white paper, Sigler spoke to the group and said, "You have seen our future."

What they had seen was the chosen strategic product focus, a major upgrading of facilities, and major changes in the organization and its culture that had been made in order to foster commitment, flexibility, and effective working relationships. It was already evident that the upgrading of both the facilities and the organization would require major investments of money and even more in time and energy. Many of the lessons and struggles ahead first encountered at Pensacola would regularly resurface in later efforts. For example

- The machine start-up was not smooth, a problem the company would continue to grapple with as it learned how better to plan and manage machine rebuilds and new machine start-ups.
- The company was grappling with the dilemmas inherent in a labor strategy that combined forcing positions on the union through hard bargaining and attempting to foster cooperative, win-win problem solving.
- Confusion and unhappiness around issues of compensation arose when a department was redesigned.

However, the strategy was beginning to take shape (as were the challenges it faced); the template was beginning to form. It was time to attempt to diffuse that emerging template to other locations, with a clear emphasis on the mills but extending to other types of work units all along the value chain from the forest to the customer.

In the spring of 1986 the then executive vice president of manufacturing, Whitey Heist, convened a one-day meeting of all the paper-mill managers to expose them to the concepts and early action steps involved in the Pensacola project. Later that summer senior executives met with a senior consultant to begin a dialogue about the future; over a period of several months and with the involvement of several layers of the organization, this dialogue led to the drafting of a statement that put the Champion Way statement into implementation terms. That statement, called "The Champion Way in Action," "reaffirms the enduring values of the Champion Way philosophy and explains how we will apply them more effectively in our daily operations by changing how we organize and manage work." It goes on to make a case for "why we must change," describes how "the way we work" must change, and lists "guiding management practices" designed to "create the environment we seek." These guiding practices include communicating fully and openly, building an atmosphere of trust, emphasizing training, delegating decision making, sharing responsibility, giving employment assurances, and creating an atmosphere that encourages innovation. The Champion Way in Action statement is presented in full in Appendix B.

In addition, a steering group called the "transition team" was established to provide direction for the diffusion efforts. Indicative of the strategic importance he placed on this effort, Sigler named Heist, who would soon succeed Longbine as president and COO, as chairman of the transition team. He held this responsibility for the decade he served as president. Assignment of hands-on responsibility for corporate change to the number two executive is rare, and it both sent a powerful signal to the organization and helped ensure sustained, tangible support for the change process. Subsequently, the transition team itself worked and reworked the Champion Way in Action statement as part of the agenda of their first several monthly meetings until it was finally ready to be rolled out.

The roll-out itself was significant in that it was consistent with the general model of diffusion used throughout the change process; all employees were exposed to the document, but each business unit and staff worked out for itself precisely which changes were to be targeted and how they were to be interpreted in order to take into account the different circumstances in each location.

The charge to the field was to take steps to start moving in the direction indicated in both statements, but, while support was available, no standard programs would be forthcoming or would be expected.

From the beginning stages of diffusion, the effort went beyond the mills. Two of the earliest redesigns were in the corporate control (accounting) and corporate technology (research and development) staffs. The successful experience in research and development (R&D) contributed later to the total redesign of all corporate technology groups, including engineering and project administration, under the leadership of Dick Olson, then in charge of all these groups. As director of corporate facilities, Tom Terfehr led an innovative redesign in that department and then served a short while as corporate vice president of organizational development. Later, as vice president of materials, he led a virtual revolution in purchasing and materials and in supplier relationships. Cross-functional teams have retooled whole processes in many areas. In addition, all support systems and structures, from information systems to reward systems, from buying to selling, have been open to reexamination, and most, if not all, have been substantively changed or affected. These are only a few examples of how change has had an impact across the value chain, as will be described in later chapters.

Support Resources

Lacking an internal organizational-development unit at the beginning of this change process, Champion turned to external consultants. The lead consultant for Pensacola was asked to take on a similar role for the companywide process and put together a team to work with each of the business units as well as the corporate staff. Private, independent external consultants, as opposed to large consulting firms, are the norm in the field of organizational development. In this case, however, individual consultants were brought together and coordinated as a team to work within a common philosophical framework toward common goals. Each worked with the mills and departments in the respective units to develop idiosyncratic change strategies tailored to the particularities of each culture and situation, yet all were guided by the vision of the

Champion Way and Champion Way in Action statements and informed by the practices adopted at Quinnesec and Pensacola; they met regularly among themselves and with corporate leaders to ensure, as Heist put it to them, that "we continue to be just one company."

One common goal of the external consulting team was to develop the capacity of each location to sustain the change process internally. Key to attaining this goal was helping each unit select, position, and train internal facilitator consultants. One year into the diffusion stage, a new position was created at the corporate level as well when a director of organizational development was named; soon after, the coordination of the external consulting team was passed from the lead consultant to him. Significantly, however, corporate leaders consistently resisted bureaucratizing this effort and were not interested in forming a large staff to drive it from headquarters. The largest this staff ever became was two: a vice president and a director; and when the second incumbent of the vice president slot was promoted to senior vice president of organizational development and human resources, he chose not to replace himself. As of this writing, the entire organizational-development staff consists of one director and two internal consultants.

A corporate group that was already in place when the diffusion effort began was an experienced and competent management-training staff, which, in the years just before this large-scale change strategy was launched, had developed a fairly sophisticated curriculum for management education and training, much of which was already consistent with the direction of the change strategy. Working with the external consultant team, they added three new course offerings: Managing Change for key change leaders, Facilitator Skills for internal facilitator consultants, and Awareness Training, which was available to all Champion employees as a way of familiarizing them with the goals and philosophy of the new strategy. Later a course called Collaborative Skills, a workshop in the basic skills of working as a team, was made available on a just-in-time basis to work groups at any level.

In addition to this training offered by the corporate staff, some mills had local trainers as well. However, with the new business strategy came new educational and training needs: training for technical operators at new facilities, cross-training for increased flexibility, training in team skills, and education in the new philosophy as well as business knowledge for employees at all levels. As the ante on training increased, mill training staffs and programs were beefed up with corporate support.

The Learning Curve: Facing the Challenges

Mills and other units proceeded in an action learning mode from 1987 to 1989 to build a foundation for change, to learn new ways of doing business, and, in some cases, even to learn the new language of the change process. This was a period of considerable ambiguity and wheel spinning, starts, false starts, and retreats. The struggles inherent in "letting go" of top-down control are covered in Chapter Four. Suffice it to say here that there were encouraging signs of success in some locations, frustrating setbacks in others, and foot-dragging in still others. A great deal of learning was going on, however, and constancy of purpose didn't waver.

No challenge was greater, though, than that faced in 1989 and 1990; Champion's mixed labor-relations strategy of forcing contractual demands (which tended to reinforce traditional arms'-length relations with union leaders) while attempting to foster effective employee empowerment and commitment came to a head when the company was faced with increasingly powerful union resistance. As bargaining took place at most of the major mills in 1989, the forces clashed over the issue of premium pay for Sundays, but this was only the tip of the iceberg of a power struggle that had been intensifying for at least ten years at Champion as well as at several other major producers in the industry. The dynamics of this struggle are covered in some detail in Chapter Three. Here it is simply important to stress that this was the most important test to date of the will and commitment of leaders of the change effort.

Many companies have retreated in the face of struggles and ambiguous results similar to those at Champion during the first five years of the change process, several falling off along the way even before the kind of union-management impasse that brought those five years to a dispirited close at Champion. The field of organizational improvement is littered with the discarded carcasses of programs that were abandoned well before they met with the difficulties of this one. When viewed from the perspective of approximately ten years, Champion's change process takes the smooth shape of the typical S-shaped learning curve found in the research on diffusion of innovation; we have overlaid this curve with some phases of the process at Champion in Figure 1.1.

During the struggle of getting the rock started up the hill, the change makers and their supporters work hard to overcome organizational inertia but with seemingly little to show for it. However, after some successful trials, diffusion begins, slowly at first, building momentum until the curve steepens as the new approach takes on a life of its own and spreads more rapidly. Finally, the change is gradually institutionalized, the pace slows, and the challenges become continuous improvement and renewal.

The reality is harder and messier than the smooth curve of the model suggests. As a middle manager at the Bucksport, Maine, mill put it, "The question of where we are in the change process is really a difficult one because one area may be way ahead of another; or, even in that area, one day you think it's going great and the next you take two steps back. Or you think you're doing the right thing and you hear people don't believe you're walking your talk. One day you're up, and the next day you're down. One minute you feel good, and the next you're frustrated. So the answer to where we are is, 'It all depends.'" Thus, for those involved—the folks "on the ground" rather than those writing about it at some distance and with a decade of hindsight—there are curves within the curve, some with such jagged edges they don't seem curvelike at all; many of the S's are truncated, starting with a flourish but leaving off in mid-air. The most commonly heard metaphor during those days in Champion was that it felt like a roller-coaster ride.

This can be, and often was in Champion's case, a fairly bumpy ride. However, conceptually, as the change process matures and the organization learns how to learn, the drop-offs are not so steep, and eventually it becomes possible to "hold the gain" at the plateau stage so that the next move forward is from that higher place on the roller-coaster track. For example, soon after the darkest days for the Champion process and after the last union contract was signed in 1990 at Courtland, two executive vice presidents

Figure 1.1. Champion's Change Process.

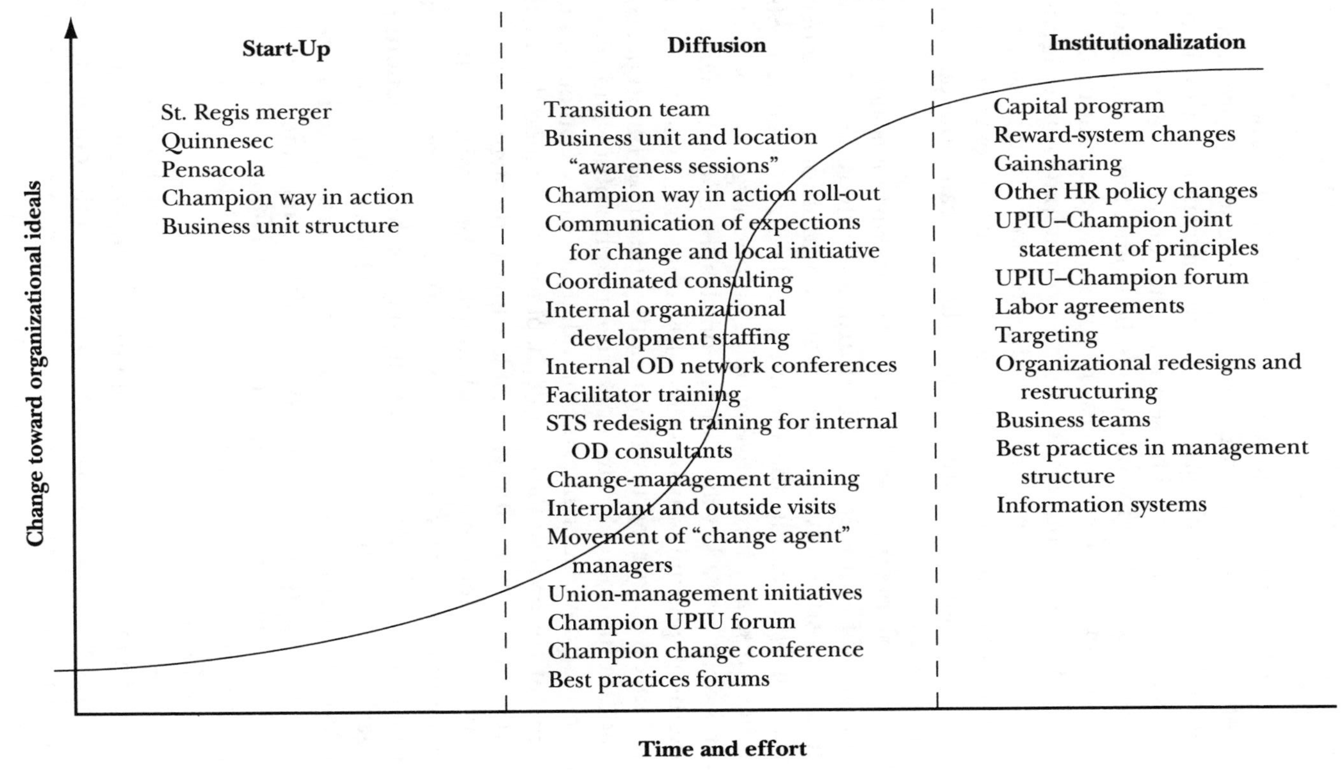

Start-Up

St. Regis merger
Quinnesec
Pensacola
Champion way in action
Business unit structure

Diffusion

Transition team
Business unit and location
 "awareness sessions"
Champion way in action roll-out
Communication of expectations
 for change and local initiative
Coordinated consulting
Internal organizational
 development staffing
Internal OD network conferences
Facilitator training
STS redesign training for internal
 OD consultants
Change-management training
Interplant and outside visits
Movement of "change agent"
 managers
Union-management initiatives
Champion UPIU forum
Champion change conference
Best practices forums

Institutionalization

Capital program
Reward-system changes
Gainsharing
Other HR policy changes
UPIU–Champion joint
 statement of principles
UPIU–Champion forum
Labor agreements
Targeting
Organizational redesigns and
 restructuring
Business teams
Best practices in management
 structure
Information systems

Time and effort

Change toward organizational ideals

independently made observations consistent with these concepts of the change curve. One took this perspective: "My theory is that, [since] 1985, we have gone up the road only about 30 percent, but it's the toughest part of the road. So all the struggle we have gone through and the millions we have spent may not all show in results but still be worth it." The second commentator seemed to sense that the next move forward would be from a higher plateau when he said, "A number of pieces are coming together at this point in time. The building blocks, such as participative management, teams, redesign, and customer-driven quality, are in place. Contracts [labor agreements] are negotiated for the next five years. We have a window of opportunity now where we can move up an order of magnitude."

With all the ups and downs there were dropouts along the way, not all of them voluntary. Unfortunately there are often some human costs involved in such massive transitions. Some people can't or won't handle the ambiguity or the stress of change. Sometimes these are supportive people. Ralph Burgess, a local union president, was a key and courageous leader of the turnaround at the Bucksport mill during 1990–1991, but eventually his committed involvement led to an unhealthy rise in his blood pressure, and, on his doctor's advice, he resigned his office and removed himself from leadership roles in other activities. Sometimes managers at any level can't or won't adapt to the new model, and, after a reasonable time with reasonable support through coaching and education, they leave either voluntarily or otherwise lest they continue to be a barrier to the progress others are prepared to make. Champion lost executives, mill managers, human relations managers, and even first-level supervisors who didn't make the shift. Union officials ran the risk of losing their elected positions because they were perceived to be too cozy with management (although the track record at Champion and other companies is generally positive in regard to the reelection of union leaders involved in joint efforts).

Champion's top leadership did not bail out, however, but stuck to its vision and strategy through these struggles and continued to find the will and resources to start back up the curve. It did not take long, however, before there was a need to face another huge challenge.

Positive Results and Industry Recession: A Tough Combination

After the final labor contracts were signed in 1990, bringing an official close to that difficult negotiations period, and as the less official lingering wounds in relationships were healed, performance at first rebounded to normal levels, as might be expected. However, it did not stop there. Now both union leaders and Champion managers realized that there had to be a better way, and they began to work together for improvement in both performance and climate. In 1991, United Paperworkers International Union (UPIU) President Wayne Glenn and Champion President Whitey Heist drafted and signed a Joint Statement of Principles regarding their desired working relationship and the UPIU-Champion Forum, a gathering of leaders from both sides, was launched in 1992. The dynamics of this changing relationship will be described in Chapter Three. The change process also matured and took hold at many other levels—in support-staff groups as well as mills, in middle and upper management as well as on the shop floor, across units and functions as well as within them—and performance curves moved upward to create record levels of quality production, uptime, and yields.

Unfortunately, at the same time a combination of new pulp- and paper-making capacity and weakened demand caused prices to drop precipitously. For example, the price drop from 1989 to 1992 for some white-paper products was in excess of 36 percent, bringing prices to 10 to 15 percent below their levels in 1986, when the change effort was launched. A drop of that magnitude can more than offset a lot of improvement in operating performance. In fact profits and operating income do not correlate closely with operating performance in this industry because of the extreme sensitivity of revenue per ton to slight shifts in demand and available pulp- and paper-making capacity. For example, Champion's operating income dropped sharply from $491 million in 1990 to $73 million in 1992 despite marked improvements in manufacturing costs and other operating-performance indices, and it improved even more dramatically from $66 million in 1993 to over $1.4 billion in 1995, out of proportion to operating improvements during this period, significant as they were. The dominant force driving these radical fluctuations was price, reflecting only moderate changes in the mix of supply and demand.

Later in this chapter, we will examine evidence of the change effort's effect on economic performance, along with other indicators, by looking underneath simple trends in operating income. At this point, in order to understand the dynamics of the change story, it is important simply to report that production rates were at record levels by 1992 and 1993 and that the cost of manufacturing, for the first time since the change process began in 1986, was lower than the industry average. Interestingly, this excellence in manufacturing competence both derived from organizational change and reinforced it. For example, Champion's people programs had helped it achieve an edge as both a low-cost and high-quality producer, which enabled it to run its paper machines full during periods of slack demand, when its high-cost competitors were forced to take downtime. An important effect of maintaining production was to provide Champion mill employees with more stable employment and earnings than was typical of the industry, a result that reinforced the evolving partnership between employees and the company.

Employee perceptions were also affected by the fact that Champion did not back off on its people efforts during this period of severe economic stress. Instead of these kinds of activities becoming the first victims of belt-tightening, as is so often the case, training continued, recruiting continued, redesign projects continued, and no pay freezes were instituted. On the other prong of the strategy, the $3.5 billion modernization program was slowed somewhat but was finally completed with the start-up of the #35 paper machine in the Courtland, Alabama, mill in 1993. All these efforts helped to convince employees as well as managers at all levels that the change process was for real, and they helped build institutional trust. Union leaders spoke of their appreciation to and respect for company leadership for sticking to its strategy during these trying times.

Thus, while the struggles of the learning curve continued, good things were happening at the same time both in measurable performance and in trust building. Nevertheless, it was a period of great frustration because bottom-line economic performance was still dismal. It was hard for employees at all levels of the company to feel upbeat let alone celebrate while losing millions of dollars. A kind of cultural inferiority complex lingered on: Are we really getting better or are we fooling ourselves? When prices return to normal—if they ever do—will we have made any gains on our competition? Despite record operating performance, there was still doubt and lack of confidence in many quarters.

Finally in mid-1994, well after the rest of the U.S. economy had strengthened, world markets—especially those in Europe—brought capacity more in line with demand, and the economic turnaround began in earnest. As prices improved rapidly and record quality production continued, Champion's success began to show up in profits and earnings. The pace also picked up in change activities, and, in anticipation of a big year in 1995, executive leadership introduced a companywide variant of employee gainsharing called Something Extra (discussed in depth in Chapter Three), intended as a one-year-only financial-incentive program to give an extra "kick" to what was hoped would be a breakthrough year.

And a breakthrough year it was. Champion achieved record sales, earnings, and free cash flow; its stock also nearly doubled as it climbed to record levels. At the same time, the American Society for Training and Development recognized Champion and the UPIU with its 1995 award for "Outstanding Achievement in Employee Involvement in the Workplace." *American Papermaker,* a prestigious industry magazine, named Champion "Company of the Year," and the title of Champion's 1995 annual report was *Good News.* In his letter to shareholders in that report, Sigler said, "Our record performance in 1995 reflected the initiatives begun nearly a decade ago, when we charted a new course for the company. . . . Our 1995 results represent an outstanding single-year achievement. What we really have focused on for the last decade, however, is building a company that will perform with maximum efficiency and competitiveness over the long term, regardless of market conditions and despite the cyclical nature of our industry. To succeed will take a strong commitment to excellence. Above all, it will take well-trained and dedicated people—people who have both the ability and desire to excel." In 1996, Champion made *Fortune* magazine's list of "America's Most Admired Companies" for the fourth consecutive year and was one of the fastest gainers from the previous year. Thus, after almost exactly a decade of transformational change—and spending over $3.5 billion in capital for new capacity (with another $2.5 billion going to projects to maintain existing capacity) and millions on developing employees and changing its culture—Champion emerged as a highly effective organization. Certainly not a quick fix; ideally "rebuilt to last."

"But," Vice Chairman Ken Nichols said in the spring of 1996, "it's a helluva business." The industry roller coaster stayed at the top only a short time, and shortly after the end of a record year, pulp and paper prices took their fastest plunge since 1974.

Results of a Decade of Change

It was against that backdrop that the new executive team took over in the middle of 1996. This team reviewed the results so far and began to reexamine strategy with a determination to build in competitive shareholder returns that would last through the down cycles as well as the good times.

A Changed Culture

In their review of the constellation of eleven pulp and paper mills, they were able to see the results of ten years of diffusion and learning. A substantial cluster of strategically important mills had made "the formula" work, including the mills in Sartell (Minnesota), Courtland (Alabama), Bucksport (Maine), Pensacola (Florida), and Quinnesec (Michigan). In all these cases, there had been extensive capital investment in the operations, and all these mills had implemented extensive systemic change over the years to support a high-commitment culture through alignment, capability, and appropriate letting go of supervisory functions—the three ingredients of the "what works" formulation to be discussed at the end of this chapter. Each engaged in work design or redesign throughout or in major portions of their operations, including the design and implementation of self-managing teams. They had used the targeting mechanism for alignment by involving employees at all levels in goal setting and planning, and some were ahead of the curve in building key performance indicators into a process for continuous improvement. Customer-driven quality was often the basis for these continuous-improvement efforts. All made use of variable compensation schemes, including pay for skills and, in four of the five, gainsharing. In the realm of union-management relations, with the exception of the nonunion operation in Quinnesec, these mills ranged from mixed but mostly positive relations to solid, mature partnerships and made widespread use of interest-based problem solving both in negotiating long-term contracts and in resolving day-to-day issues. All were also marked by a strong commitment to continuous education and training.

Again it must be emphasized that change didn't go on in just the paper mills. Several corporate support groups (such as applied technology, control and accounting, materials, facilities, credit management, and information technology) could be clustered with this group of mills both in degree of change and in performance. Moreover, the forest-products business unit as a whole had been a leader in the change process for the entire company, with, in 1997, such forest-resources (timberlands) locations as the Lakes States (in Upper Michigan) and Western Florida as particularly exemplary cases. Additionally, Champion's Weldwood subsidiary in Canada launched an ambitious and comprehensive change effort in 1995 called the Weldwood Commitment, while the operations in Brazil had always been among the most successful and people-oriented in the company.

In addition to the five high-performing domestic pulp and paper mills cited above, two other mills had not made such extensive systemic changes and had not benefited from major capital investment, but, by building on inherently sound relationships with proud, skilled employees, they had performed well despite the odds. The mill at Roanoke Rapids, North Carolina, was isolated from the strategic mainstream of Champion, producing specialty packaging products on old machines for a market far different from commodity papers. However, with solid alignment of their total organization to their customers' needs, cooperative labor relations, and a tradition of competence, this mill regularly made a profit and had one of the best safety records in the company. To reinforce this alignment they had also recently introduced gainsharing. The mill in Lufkin, Texas, also had to make do with old facilities because nothing short of a huge investment in starting over from scratch could have brought the technology to modern, competitive standards. This mill was also in the problematic newsprint market, a segment of the paper industry in which Champion found it especially difficult to make a profit. Nevertheless, with a solid workforce and great emphasis on building cooperative union-management relationships, including joint union-management participation in targeting and continuous improvement, the plant had improved.

The remaining mills had sputtered along over the ten years in both the change process and performance. The Canton (North Carolina), Deferiet (New York), and Sheldon (Texas) operations all had problems with finding profitable markets for the products they could produce given their technologies. In regard to the change process, each had experienced ups and downs, having a period or two in the limelight when it looked as though they were making real progress but then seeing their efforts trail off, incomplete—a pattern much like typical change efforts in many other companies. Champion's original home mill in Hamilton, Ohio, fit this pattern for most of the decade; however, in the last two years mill leadership undertook a total transformation process with the goal of becoming a high-performing, highly profitable specialty-papers mill. Major downsizing had taken place with union involvement and cooperation, significant capital investment was made, and there was a commitment to a totally participative redesign of work systems across the mill. The jury is still out on where these changes will eventually place Hamilton in the array of mills just described.

The amount of change represented by this array is significant. For example, as shown in Figure 1.2, the percentage of employees whose jobs changed through redesign ranged from 30 percent at the low end to about 80 percent at Sartell and in the forest-products business unit. Although many companies can point to successful efforts at one or a few locations, few have so many of their core, strategic operations so far along on the change curve. In addition, as the senior team reflected on the amount of change in both culture and performance at the location and staff level, they also had to acknowledge that, at the corporate-system level, the change was greater than the sum of the parts. Champion as a whole was a far different company than it had been in the mid-1980s.

**Figure 1.2.   Percentage of Employees Whose Jobs Have Had
Role or Redesign Changes.**

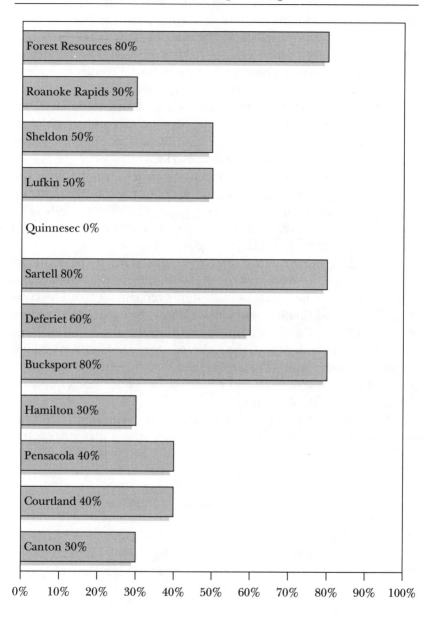

Improved Performance

We examined an abundance of evidence to assess the effects of the change effort on economic and operating performance as well as on social indicators. As mentioned earlier, because of the profound influence of price (which is largely uncontrollable) on economic performance, we must look underneath simple trends in operating income into two general areas. First, we look at changes in operating performance over the decade, including manufacturing costs, productivity, quality, and waste. Second, we compare changes in economic-performance indicators at Champion and other paper companies producing for the same markets.

In terms of operating performance within Champion over the decade, mill production increased by 32 percent and employee headcount decreased by 12 percent from 1986 to 1996 (Figure 1.3), a productivity increase of 47 percent as measured in annual tons per mill employee (Figure 1.4).

Operating-performance improvement was not just a matter of cranking out more tons however. Champion's product quality and customer service have dramatically improved since 1986. The company has earned quality awards from Xerox, Sears, Moore Business Forms, Tension Envelope, and *Reader's Digest*, among others (see box). No other company in the industry has come close to this kind of recognition. Innovations brought about through best-practices efforts transferred directly into economic benefits; these efforts will be covered in Chapter Four.

In assessing Champion's operating performance versus the performance of its competitors, one of the most instructive

Xerox	"Award of Excellence"
Sears	"Partners in Progress"
Avon	"Supplier Excellence Award"
Reader's Digest	"Supplier of the Year"
Tension Envelope	"Outstanding Supplier of the Year"
Moore Business Forms	"Supplier of the Year Award"
Data Documents	"Vendor of the Year"
R. J. Reynolds	"Preferred Supplier"
Mead	"'A' Rating"

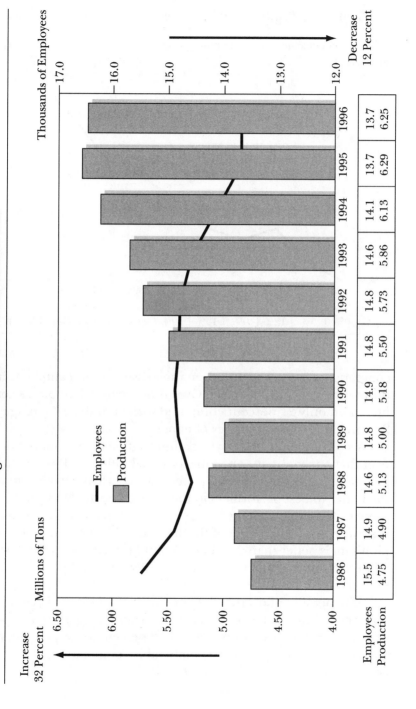

Figure 1.3. Production Versus Headcount.

	1986	1987	1988	1989	1990	1991	1992	1993	1994	1995	1996
Employees	15.5	14.9	14.6	14.8	14.9	14.8	14.8	14.6	14.1	13.7	13.7
Production	4.75	4.90	5.13	5.00	5.18	5.50	5.73	5.86	6.13	6.29	6.25

Figure 1.4. Productivity Improvement per Mill Employee.

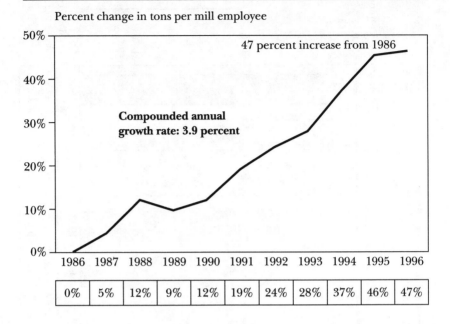

Percent change in tons per mill employee

1986	1987	1988	1989	1990	1991	1992	1993	1994	1995	1996
0%	5%	12%	9%	12%	19%	24%	28%	37%	46%	47%

47 percent increase from 1986

Compounded annual growth rate: 3.9 percent

comparisons focuses on trends in their cash cost of manufacturing during the period 1986–1996. Over those years, Champion's costs increased only 21 percent compared with an industry average of 25 percent for similar grades of paper.

According to Jaakko Poyry, an international and independent firm that analyzes paper producers' capabilities, by 1995 virtually all Champion's pulp mills and paper machines had become among the better to best performers in the world. The analyst found that some of Champion's mills had achieved uptime and paper losses that were better than state of the art. It rated 10 percent of Champion's mills better than the "Better/Best Performers" in the world; 87 percent were equal to the "Better/Best," while 3 percent were poorer.

As a measure of overhead, between 1986 and 1996, Champion reduced selling, general, and administrative expenses as a percentage of cost of sales from about 10.8 percent to 7.8 percent, the lowest level among eight most comparable competitors, whose percentages in 1996 ranged from 8.1 to 10.1.

Measures of safety serve as both economic and social indicators. Not only has Champion improved its safety record since 1986, but its improvement has been much more dramatic than the industry average. As can be seen in Figure 1.5, from 1986 to 1996, Champion cut its injury rate by nearly two-thirds, whereas the industry average dropped by only less than one-third. The improvement at Champion resulted in workers' compensation savings of about $15 million, but more important it made Champion a much safer place for people to work.

Some developments in the union-management relationship may also serve as social indicators. For example, the length of union contracts expanded from a routine three years to a standard of five, with one contract for seven years. Moreover, while the earlier contracts were bargained in a traditional adversarial way, the latter, longer agreements were largely worked out using nonadversarial problem-solving processes in a win-win atmosphere. As an outside indicator, the previously mentioned 1995 award from the American Society for Training and Development to the company and union for "Outstanding Achievement in Employee Involvement in the Workplace" included the following words of commendation: "there are few (if any) other union-management efforts in the U.S. which have been so strategic (top down, tied to strategy); pervasive (wide-spread throughout the company); persistent (long-term, on-going); and successful."

Perhaps the yield from the decade of change was best captured, however, by Nichols when, in 1995, he presented to Champion management and employees one illustration of the earnings impact from running the mills better. He took 1992 as the base year because it came well after the effects of the bargaining struggles of 1989–1990 and because it was itself, for that time, a year of record production. The only new capacity resulting from a capital project during the period from 1992 to 1995 was the addition of the #35 paper machine at Courtland, and those additional tons were subtracted from the comparison. The rest of the improvement he called improvement from running the mills better, including all the change initiatives and everything else that people did to contribute to continuous improvement. Running the mills better resulted in an increase of 395,000 tons over the three years. Most important, that increase had an impact of approximately $1.42 on 1995

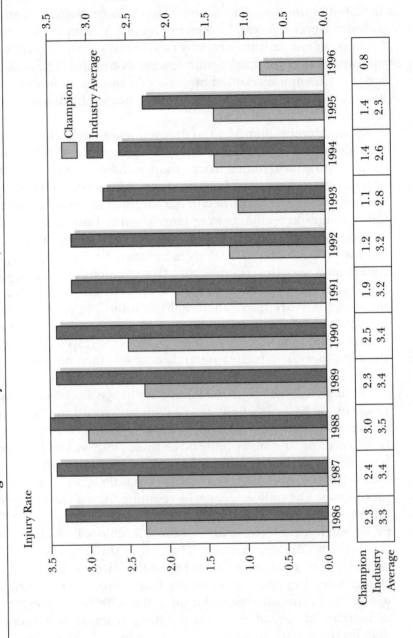

Figure 1.5. Safety Trend: Lost Workday Case Injury Rate.

Injury Rate

	1986	1987	1988	1989	1990	1991	1992	1993	1994	1995	1996
Champion	2.3	2.4	3.0	2.3	2.5	1.9	1.2	1.1	1.4	1.4	0.8
Industry Average	3.3	3.4	3.5	3.4	3.4	3.2	3.2	2.8	2.6	2.3	

earnings per share, fully 17.7 percent of record earnings. Dramatically impressive to papermakers, however, was his conclusion that the result of running the mills better was comparable to having added one and one-half new paper machines of the size and capacity of Courtland's state-of-the-art #35 without having spent the capital, which in 1995 costs of construction would have amounted to an expenditure of $650 to $700 million. In addition to this enormous savings in capital expense it was worth far more in bottom-line savings through incremental increases in tons of production.

Distinctive Factors in Champion's Change Strategy

Over the decade, Champion spent a lot of money and engaged in a great number and variety of initiatives while it struggled with daunting outside forces, its own organizational inertia, and its own learning curve. Many of its initiatives were not unlike those in other companies in the United States or, for that matter, in the world. At the heart of the Champion story, however, and what distinguishes it is that the change strategy was both sustained over a long period of time and comprehensive in scope. These two factors make this story special.

Sustained Change

It is conventional wisdom that U.S. business leaders today tend to emphasize the current quarter's earnings, constraining their decision making by such limited measures and time horizons. When it comes to "improvement initiatives," the landscape of many U.S. corporations is littered with short-lived programs of the past, faddish flurries of activity that quickly disappear when the next fad comes along. Employees subjected to the "program of the month" become numb and develop "programmitis." Whether or not these observations represent the generalizable "truth" about U.S. enterprise, it is clear that this sort of myopia is inimical to fundamental organizational change.

Harvard's Michael Beer and his colleagues found in their research into successful change efforts that "companies need a particular mind-set for managing change: one that emphasizes process over specific content, recognizes organization change as a unit-by-unit learning process rather than a series of programs, and acknowledges the payoffs that result from persistence over a long period of time as opposed to quick fixes. This mind-set is difficult to maintain in an environment that presses for quarterly earnings, but we believe it is the only approach that will bring about successful renewal."2 Champion executive vice president Dick Porterfield put it more succinctly when he observed that what was required was "dogged persistence and unjustified optimism."

Neither were lacking at Champion, even during the most difficult and troubling times, and much of the credit for that must go to the leadership provided by Sigler. From the beginning he was pressured by Wall Street for short-term results for shareholders. He was criticized for spending billions on building and rebuilding pulp- and paper-making facilities. After emphasizing the people side of the process beginning in 1986–1987, he saw mill operating performance actually nosedive along with employee attitudes during labor-contract bargaining in 1989–1990, as described above. Then, as soon as contract disputes were resolved and mill performance began to rebound, the industry was hit by a prolonged and serious down cycle. Presumably Sigler and Heist had their private doubts from time to time, but outwardly they remained steadfast and confident in the ultimate wisdom of the vision and strategy.

As concrete evidence of what one outside observer called the "indefatigable will" at work, they also put their money where their mouths were. In the depths of the recession of the early nineties, when it would be normal for companies to take such belt-tightening measures as downsizing, layoffs, and pay freezes, Sigler strongly maintained that such temporary squeezes don't work. "I should know," he said. "I've tried them all at some time or other" in his long term at the helm. His position was that these kinds of actions only lead to a rebound effect later on that makes for higher costs and less efficient operations. "You cannot save yourself into prosperity in this industry," he insisted. "You have to perform better than anyone else." So those factors that contribute to performance—investment in state-of-the-art facilities and the people to run them—went forward.

By this time, some of the change had begun to take hold, and Sigler and others believed that the results they were seeing with the company's people efforts gave Champion a competitive advantage that would become visible to the world when the economy turned around and reasonable capacity-demand balance was restored. Indeed, after the labor conflict ended but still during the downturn, mill performance was at record levels and continuously improving, hidden to outsiders only by severely depressed paper prices but solid enough to reinforce the decision to continue recruiting, training, and fairly compensating people, developing effective union-management relations, and engaging in the extensive array of initiatives aimed at alignment, capability, and letting go of direct control. This steadfastness contributed as nothing else to the constancy of purpose and the degree of clear alignment that evolved in that decade.

Comprehensive Change

The second distinguishing characteristic of Champion's change effort is its comprehensive scope. Almost every approach Champion experimented with or implemented has been taken elsewhere. A number of organizations have now been involved, more or less successfully, in implementing such "shop-floor" initiatives as sociotechnical redesign, self-managing work teams, pay for skills and contribution, competency-based training, and statistical process control. Although these initiatives are important parts of Champion's successful change effort as well, they are not what separates the Champion story from the rest.

Champion's change strategy has been comprehensive in scope and activity from the beginning and throughout the decade of change. Indeed, we know of no other company in the world or in the history of efforts of this nature to do all of the following: approach broad technical and social-system change as central to the firm's fundamental business strategy; effect the change horizontally across the value chain from suppliers to customers, including the crossing of internal organizational boundaries; involve all parts of the organization and all levels of the hierarchy; and include substantive changes in policies, systems, and other infrastructure at the corporate level. In addition to this breadth, the depth of change has truly been profound, as groups at all levels and in many different functions have come to see the business and their roles in it in a fundamentally different way.

Some Actionable Ideas

What lessons already emerge from our telling of the Champion story? Each reader will make his or her own sense of the story, but here are a few ideas we take away from this chapter.

Contributing critically to the remarkably sustained nature of the change effort were the following:

- The fact that management did not rely on a crisis or urgent threat to mobilize change. Rather it took advantage of events that presented opportunities for dramatic innovation in certain locations—for example, the greenfield mill at Quinnesec and the newly acquired Pensacola mill, which had already entered into an "unfrozen" state.
- The continuity of leadership at the top—including Sigler and Heist, the executives who replaced them, and others now on the top management team.
- The "dogged persistence and unjustified optimism" along with the common commitment that these members of top management shared. Much managerial time and many organizational resources had to be invested over a considerable period of time before results could provide hard evidence of the success of the change effort, especially inasmuch as the severe price erosion created worsening financial results. Sustained financial support for the change process through both good and poor economic periods helped reassure the organization that the change process was integral to the business strategy and not a program that could be indulged in only in good times or that was seen only as a remedy for bad times.
- The early emphasis of change leaders on human and organizational-development goals followed by increasingly demanding aspirations for quality and economic performance. The sequence worked well in establishing the change agenda and continually challenging and stretching the organization to achieve goals not previously contemplated.

The unusually comprehensive nature of the change effort appears to be due to many factors including importantly the following:

- The top executive (while himself maintaining continuing involvement) placed the responsibility for the change effort directly in the hands of the number two executive, thereby ensuring both that change had top priority and that it was seen by others in the organization in the right light. The number two executive created a team of key players to guide the transition process.

- Every top functional executive participated in developing the Champion Way in Action. Their participation promoted the view that the statement applied to employees in all parts of the corporation, including professionals and managers as well as production workers. The top group appeared to believe that "what is good for the goose is good for the gander," although they never articulated this belief directly. We have observed that many other corporations encounter great resistance from staff groups when changes already implemented in production operations are urged on them.

- The transition team adopted a combination of (1) pressure on mills and other operating units to produce changes in line with Quinnesec and Pensacola while providing (2) latitude for units to choose their own specific change targets and processes and (3) resources for units to employ consultants to help in the change process.

A Conceptual Framework for Managing Change

These seven factors are primarily about the strength of corporate leadership and the general stance it took toward the change effort. How leaders at all levels managed the change process is the subject of the next three chapters. The specific actions leaders took—such as articulating goals, convening change forums, providing training, involving additional people in decision making, delegating authority and responsibility, and redesigning jobs and structures and systems—are organized and examined according to the conceptual framework we developed from our interviews with Champion employees at all levels within the organization (see box).

The framework we developed and present here is the way *we* made sense out of the change history after the fact; however, we believe it has general validity. Although it is not a framework we developed before we began to analyze Champion's experience, it is one we now find helpful in examining other cases. And

The following quotes are representative of the responses of Champion employees who were asked, "What works and why?"

- Good people who see their future is up to them; good information widely shared about both the "whats" and the "whys"; freedom and power to act to do something about the situation. (Lufkin, Texas, paper-mill superintendent)
- Freedom to do your job the way you know how. (Corrigan, Texas, plywood-mill hourly worker)
- A clear, specific, worthy goal toward which everyone pulls together and doesn't worry about status or anything else. (Canton, North Carolina, paper-mill clerical worker)
- This mill runs best when people are left to do their jobs. When that happens, everybody pulls together, helps each other, works as a team, without concern for whether or not it's their job. Supervision is available for help when needed, but don't try to tell long-term, experienced people how to do their jobs. (Roanoke Rapids, North Carolina, liner-board-mill union member)
- Common understanding and ownership through open communications, with freedom to act as part of a team to improve the situation and be recognized for your contribution. (Pensacola, Florida, paper-mill department head)

although it was not used by Champion managers to guide their change initiatives, we do now recommend it to others who are attempting to manage systemwide change. Indeed, it is precisely because we believe that it helps us capture a generalizable set of lessons that we use it to tell the Champion story.

We express this framework as a "formula":

$$W^2 = A \times C \times L$$

- W^2: *what works.* We have reported in this chapter some of the evidence that the change process is indeed working.
- *A:* the *clear alignment* of the organization and the organizational change effort with business *strategy, values,* and core *technology.* In Chapter Two we analyze steps taken consistent with these three touchstones of alignment.

- We have good people, and when they have good communications and training and lean management, they can get a good job done. (Sheldon, Texas, paper-mill hourly worker)
- This place works well when three things are present: (1) we overcome the "it's good enough" mentality; (2) we do our up-front planning and communications well; and (3) we tap into the we-can-do-it-and-we-want-to-do-it spirit of the people here. (Deferiet, New York, paper-mill manager)
- Decisive leadership providing clear direction with full communication of the organizational consequences, combined with our people's willingness to work without formal, bureaucratic boundaries and rules. (Quinnesec, Michigan, paper-mill department head)
- Providing people with all the information they need to act individually or as a team and then taking the risk to let them loose to act. (Texas Region forest-resources middle manager)
- When people get aligned with a purpose they understand and believe in and then management lets them go. (Sartell, Minnesota, paper-mill organizational-development internal consultant)

- *C:* the *strong capabilities* that result from skilled, dedicated people working together. In Chapter Three we refer to these factors as the Three Cs of Capability—*competence, commitment,* and *cooperation*—and we describe some of the ways they were fostered at Champion.
- *L: letting go* appropriately of direct supervisory control (or providing optimum latitude). Letting go is a simple concept, but it is difficult in practice in part because it's not a case of "the more the better" but of balance and appropriateness in the specific circumstances. Chapter Four covers how Champion applied the Three Ss—*systems, structure,* and *style*—to address these dilemmas of letting go.

Why these three ingredients? Each is necessary and together they are sufficient to produce effective change. Each complements the other two. Letting go—empowering subordinates—creates energy and initiative. Building capability means harnessing the energy for employee and unit performance. Achieving refined alignment of the firm's strategy, task technology, and social values ensures that employee and unit performance is properly directed for competitive advantage and is not at the expense of other purposes, such as social ideals, to which the firm is committed. Throughout our discussion of these three ingredients we also emphasize the interaction among them—how, for example, the clearer the alignment of members of the organization with strategy, values, and core technology, the easier and more effective it is to let go of closely supervising their work.

Table 1.1 presents a relatively comprehensive list of the managerial levers used by Champion in stimulating and managing change. It is organized vertically by conventional categories of managerial levers, while the rows across the table indicate how the levers are related to the conceptual framework and in which chapter or chapters each lever is discussed.

Table 1.1. Managerial Levers Used to Create Alignment, Build Capability, and "Let Go" of Direct Control.

Managerial Levers	To Create Alignment (Chapter Two)	To Build Capability (Chapter Three)	To Let Go (Chapter Four)
Job and structural redesign			
Mills	•	•	
Accounting	•		
Corporate technology	•		
Interfunctional teams	•	•	
Management information systems			•
Top management			
Corporate structure	•	•	•
Rewards	•	•	
Pay for skills		•	
Gainsharing	•	•	
Something extra	•	•	
Bonuses		•	
Recognition		•	

Table 1.1. Managerial Levers (continued).

Managerial Levers	To Create Alignment (Chapter Two)	To Build Capability (Chapter Three)	To Let Go (Chapter Four)
Participation mechanisms			
Off-site sessions	•		•
Teams	•	•	•
Goal setting, planning, and information policies			
Targeting	•	•	
Open-books management	•		
Human resources policies			
Training and education	•	•	•
Selection			•
Appraisal			•
Employment security		•	
Union-management initiatives			
Fostering tactics		•	
Forcing tactics		•	
UPIU–Champion forum		•	
Joint statement of principles		•	
Best-practices initiatives			•
Leadership style			•
Project management	•		
Change-process structures			
Transition team	•		•
Lead consultant	•		
Internal organizational-development staffing and training	•		•
Language: banners for change	•		
"Participative management"	•		
"Redesign"	•		
"Customer-driven quality"	•		

Chapter Two

Aligning the Organization
Strategy, Technology, and Values

We are struggling to juggle three top priorities—the capital-rebuild project, operating performance, and the change process. Each has its Stamford [headquarters] advocates who are pushing their priority. Just as it would be helpful if we got clear ourselves on how to manage the three, it would be helpful if they would be better coordinated themselves.
—Mill middle manager, October 1991

Our profits are much improved, but that's not really our focus at all. It's not the reason we do the things we're doing. Our focus is on our customers—the Quinnesec and Sartell mills. All of us are totally aligned to meet their needs. Our profits are a by-product of that.
—Forester, August 1992

In discussing the idea of *alignment,* different employees cite different examples. One may point to the need to meet Xerox quality standards; a second may refer to achieving tighter coordination between a paper machine and its downstream converting operations; a third may stress the importance of involving people who are going to be affected by a decision. And each of them is correct—actions need to be aligned in all three ways: with competitive strategy, with key links in the core technology, and with social values. In this chapter, we will focus on the alignment of

Figure 2.1. Three Forces.

these three touchstones with Champion's change process and specific change initiatives with those three forces represented in the model shown in Figure 2.1.

The basic proposition advanced here and reinforced by Champion's experience is that the better the change effort and change target are aligned with a firm's competitive business strategy, enduring social values, and core technology, the more effective the change effort will be and the more powerfully it will contribute to the kinds of business and social performance highlighted in Chapter One. In this chapter we will illustrate a number and variety of initiatives designed to align actions and decisions of individuals and groups with one or more of the three touchstones. We will also examine how these alignment activities (and the explicit attention to alignment itself) evolve over time and as the focus of attention moves up and down the organizational hierarchy. Finally, we will show how alignment affects the other two ingredients in the performance formula—capabilities and letting go—as shown in Figure 2.2.

Alignment Touchstones

If the change process is to be aligned with competitive strategy, social values, and core technology, what do we need to know about the nature of these touchstones in Champion's case? How and when were they clarified so that they could be used to shape the vision itself and align the change effort? Sometime in the mid-1980s

Figure 2.2. The Performance Formula: Alignment.

Alignment	× Capability × Letting Go = What Works
• Competitive strategy	
• Social values	
• Core technology	

CEO Andy Sigler and his immediate colleagues began to envision a different type of corporation. This vision was driven by competitive considerations and technological imperatives and was also shaped by social values. This combination is an important part of the explanation for the soundness of the evolving vision based on the Champion Way in Action statement and the robustness of the change effort it spawned.

These are the most fundamental forms of alignment necessary for effective transformation. If the idealized organization is not tailored to the economic realities of the marketplace and its particular production technology, it won't have maximum leverage on the key competitive success factors. Similarly, if the idealized organization is not consistent with fundamental assumptions about human motivation and with values regarding how the corporation should conduct itself, the vision will neither capture the imagination of employees nor be reliably implemented.

Competitive Strategy and Technology

After the merger with St. Regis in 1984, the company made a strategic decision to focus on the white-paper commodity businesses, shedding both the brown-paper business and many other unrelated businesses. By 1986 the corporation was composed of (a) eleven mills committed to several major businesses—printing and writing papers, publication papers, newsprint, and pulp; (b) approximately 5.5 million acres of timberland under management; (c) plywood and stud mills; and (d) a few downstream businesses, such as a paper-distribution company.

By far the bulk of the corporation's assets and earning potential was concentrated in the production of pulp and paper for white papers. The company sought to become a low-cost producer of white commodity papers with as high standards of quality and responsiveness as its best competitors and to do this by two means: first, by instituting a massive modernization of existing mills, increasing tonnage capacity, increasing reliability, and lowering variable costs per ton; and, second, by operating and supporting these mills with the most effective organization in the industry.

These two prongs to the competitive strategy were dictated in part by the embedded nature of the infrastructure of pulp- and paper-production technology. Papermaking plant and equipment are both expensive and fixed in place; therefore, unlike the case in many other industries, a paper company can't readily shut down a mill and start one in Mexico or elsewhere in order to hire a new workforce or acquire all new technology. Champion had to work with what it had in both mill locations and workforces.

The nature of the production process also created the high economic leverage of organizational effectiveness and determined what characteristics made the organization effective. First, pulp and paper production is highly capital intensive (a billion-dollar mill can be operated by fewer than five hundred employees). Therefore, to achieve low costs it is extremely important to keep equipment running, which requires a smart, committed, informed workforce with strong problem-solving capabilities. Second, the cost of raw materials is a significant fraction of the cost of goods, underscoring the importance of achieving good yields. Achieving high yields depends on having a workforce that cares about minimizing waste in a complex process and knows how to do so. Third, although control technology for pulpmaking and papermaking has advanced in recent decades, there is still a lot of art left in the production process. The uncertainty in the production process derives in part from the unavoidable variability in raw materials and the tight interdependence of all the steps in a continuous-flow process. Thus, the economic performance of a mill is especially sensitive to the commitment and competence of the workforce as well as to their ability and willingness to work as a team within and across their crews.

The commodity nature of the white-paper business also reinforces the type of attention given by Champion to production technology and human organization. Neither Champion nor any of its competitors dominates the markets in which it participates. Moreover, in this business, there are no magic bullets in terms of new-product development or innovative advertising campaigns. One wins only by perfecting each and every production, maintenance, sales, and administrative activity. Therefore, excellence in operating the mills themselves is a necessary condition for competitive advantage but not a sufficient one. Champion management concluded that corporatewide change—from one end of the value chain to the other and from bottom to top—is the only way to win day in and day out.

Finally, Champion management gradually realized that inasmuch as all these conditions applied equally to their direct competitors, Champion would need to drive the change process more boldly, more effectively, and more persistently than any other paper producer.

Social Values

If economic forces and the nature of the production technology provided Champion management in the mid-1980s with a strong incentive for generating a motivated, competent, and high-performing workforce, a value statement already formulated provided guidance for management's approach to creating such a workforce. Certainly the development and application of the Champion Way and Champion Way in Action statements, covered in Chapter One, were important alignment initiatives. In the first year or two of the change process, these documents were discussed, pulled apart, examined, tested, and challenged in every way imaginable. As time went on and the process matured, they became not so much a topic of daily conversation as a solid foundation for any and all actions and initiatives. By no means, however, were they hung on the wall or put in a drawer and forgotten, as often happens. Their pervasive influence throughout the company was a clear demonstration that top management would allow for a great deal of freedom of action within the limits of highly valued and deeply felt principles.

Aspiration Levels

Each of the touchstones was ambitious in its own right and was based on high underlying aspirations. As the Champion Way in Action statement, presented in Appendix B, stated publicly, the economic goal toward which the competitive strategy was aimed early in the change effort was to put Champion's profitability in the top quartile of U.S. industry—a difficult but not unprecedented accomplishment for companies in the commodity paper industry. The technology envisioned by management was certainly not their existing equipment; rather it was the best affordable and available technology. Similarly, the idealized social values were based on highly optimistic assumptions about human behavior and high aspirations for developing and tapping human potential.

Ambitious goals—providing they are not unrealistic—drive the level of achievement. The trick is to raise aspirations repeatedly in conjunction with demonstrating achievements on many fronts: quality, internal relations, union-management relations, financial performance, and so on. And one of the arts of management is to find levers for jacking up aspirations and standards—levers that are legitimate and effective. Champion management attempted to do this by a variety of means—a theme that cuts across the rest of this chapter and Chapters Three and Four.

Aligning Change Initiatives: The Transition Team

A main characteristic of the Champion change effort as summarized in Chapter One was its sustained and comprehensive nature. A fundamental challenge in managing hundreds of individual change initiatives—at different places and at different times—was to achieve consistency among them. A new initiative had not only to be consistent with the three touchstones but also to build on and reinforce the existing initiatives.

In late 1986, as the change process began to be intentionally diffused from its early manifestations in Quinnesec and Pensacola, it was recognized that some sort of central steering, support, and integrating mechanism might be necessary. Although the change-management philosophy was to encourage each unit and location to develop in a manner consistent with its own unique culture and state of readiness, it was also believed to be important that learning take place across units and that some overall coordination was needed. The mechanism created for this purpose was the "transition team," chartered as a forum for learning, for monitoring the change process, and, in the team's words, for "keeping the flame alive." The group consisted of the executive vice presidents of each of the four business units; their manufacturing vice presidents; the staff heads of accounting, manufacturing support (which included R&D and engineering), and human resources; plus "support resources" from internal communication, employee relations, training, and management development. It was headed by the COO, President Whitey Heist. Sigler was an ex-officio member but ended up attending most of the monthly meetings.

The power of this membership roster was a further demonstration of the integrity of the company's commitment to this strategy; in many such efforts, the "steering committee" is not nearly so strong. On occasion the whole group visited other companies' plants to learn from what they were doing, and, inevitably, the leaders of those plants expressed their envy, saying that they wished their company provided the same understanding and support at the top.

This group met regularly, seldom missing a month, from the last quarter of 1986 to 1992, exchanging experiences, visiting locations in and out of Champion, learning from outside experts and practitioners, and assessing progress. Although sometimes the meetings consisted of predictable reports from the field, at their best they were, as one member put it, "like attending a seminar each month."

The transition team was not a decision-making group; rather, based on the shared information and learning, the executives involved would make decisions in their respective organizational roles through their business units or other normal decision-making bodies. As the new Champion matured, the transition-team activities were carried out by a reorganized top-management group known as the "Treetops" group.

Aligning Mind-Sets

It is not sufficient for organizational initiatives to be aligned; all those who are affected by and can affect the initiatives need a common way of thinking about them and their rationale. In brief, the organization should strive for a common mind-set about the change.

At Champion three major devices for aligning mind-sets were the language used to label the change effort; the models used to clarify how aspects of the change effort fit together; and the orientation, training, and development sessions held to discuss it. We will explore important illustrations of each of these devices.

An Evolving Set of Banners

Language creates context, and alignment is a matter of creating a common context; therefore the language used to communicate about the change process can be crucial. Choosing an appropriate label for an improvement process is difficult however. Many of the labels often given to such efforts limit the process to one approach or method, or they're faddish buzzwords or consultant jargon; others put the emphasis on means rather than ends; while others merely add ambiguity to an already fuzzy process. Probably there can be no absolutely correct label; any label can get in the way. However, people need to have some language they can use to talk about "it." Champion did not avoid these difficulties with labels.

The process was called different things at different times. In the early stages, it was most commonly called *Participative Management,* or *PM,* although this was not an "officially sanctioned" title, and no one either in the company or among its external consultants claims to have first applied the term to Champion's efforts. This label did seem to coincide with a *perceived* emphasis on relationships. No matter how much change leaders or consultants spoke of "harder" interventions, such as change of structures and systems or improved operating performance, what people seemed to hear in the early days was "improve working relationships." We have observed the same phenomenon in other cases; it may indicate that people feel the need to address dysfunctional relationships before they can see past them to some of the problems or goals perceived as less "soft." As late as 1991 people asked, "How's your PM effort?"

Redesign was another early label, although not so pervasive. It referred to the redesign of both individual jobs and work organization. Thus, it too emphasized a means rather than an end, and a limited means at that. There was, in fact, an early misperception by some that sociotechnical redesign was the one and only officially sanctioned change methodology, perhaps because redesign was a key feature of the change process at Pensacola, the first established mill to be targeted for change.

Concepts and tools for managing quality had a good deal of currency nationally during the decade of Champion's change effort. Consistent with Champion's general stance of not committing to any one orthodoxy while considering all promising approaches, these ideas and methods were explored and selectively applied in an integrative way with other initiatives while at the same time leaders attempted to avoid faddishness and the creation of another "new" and bureaucratic program. However the fundamental concepts of customer focus and continuous improvement were wholeheartedly adopted in 1989, after which the effort was often referred to as *Customer Driven Quality*. Compared with both Participative Management and Redesign, this label was considered to be oriented more toward performance and end goals and aligned with business strategy rather than oriented toward participation for participation's sake or relationships for relationships' sake. On the downside, however, many people perceived Customer Driven Quality to be a "new program"; as a result, they didn't relate to it as part of the unbroken, continuous process that had started with Participative Management and Redesign and that included such other approaches as competency-based training and targeting (a goal-setting and accountability system, described below).

Interestingly, although not by conscious design, the three labels that had strong currency at different points in time each emphasized a different touchstone. Thus, Participative Management referred to *social beliefs* that employees have much more to contribute than they traditionally have been allowed to contribute and that they deserve an opportunity to do so. As this social idea became widely accepted, Redesign—and the sociotechnical method associated with it—emphasized the need to organize in a way that optimized not only general social criteria but also features of the *core technology*. Then, Customer Driven Quality subsumed the earlier ideas in a concept that, as mentioned above, was aimed at an important aspect of *competitive strategy*. In hindsight the sequence of banners was fortuitous.

Throughout all this time, some had referred to everything going on as simply the *change effort* or the *change process,* and, eventually, almost everyone used these terms as general umbrellas to cover all the initiatives that were part of changing and improving the company following the merger with St. Regis. What to call "it" became less and less of an issue. This last phase in the evolution of terms was in itself symbolic of the maturation of the change process.

A New Model

It is useful to delve into the role of one of these banners to see how they can align mind-sets. Customer Driven Quality became a helpful banner because the focus on customers and quality provided a strong rallying point around which to build consensus and commitment, and it targeted the change effort on the strategic touchstone. When the phrase *Customer Driven Quality* entered the change-process vocabulary in 1989, a new model was developed that demonstrated both the scope and the alignment of the initiatives under way (Figure 2.3). Quality as a concept is endorsed by everybody in all organizations, at least in words if not in behavior. As such, it represents a common ground. However, internal channels of communication about quality based on internal specifications are contaminated with a lot of "noise." "Can I trust the information from management? Workers? What's their angle? Why are they telling me this? I don't like him/her anyway, so what do I care? That spec isn't really important anyway." However, it is not too difficult for everyone, regardless of position or status in the company, to understand that both individual and organizational security depend on customers, and whether you trust or like them or not, when they speak, it makes sense to listen. Direct communications channels with customers are much cleaner, and when the communications focus on quality, both channel and message promote alignment.

For example, union and management at the mill in Bucksport, Maine, formerly one of the prized mills at St. Regis, were at loggerheads in the early years of the change process. A fairly common feeling among employees was that the new owners, Champion, didn't like them and they didn't like Champion. These feelings

Figure 2.3. Model Based on Customer Driven Quality.

- •Supplier partnering
- •Benchmarking
- •Cost
- •Statistical process control
- •Operator skills training
- •Site visits to customers
- •Customer teams
- •Identify key customers
- •Quality-cost relationship
- •Define customer requirements
- •Targeting

- •Flexible benefits
- •Targeting
- •Nontraditional compensation
- •"Line" organizational structure

Process improvement

Leadership

Customer focus

Workforce empowerment

- •Work redesign
- •Targeting
- •Team building
- •Nontraditional bargaining
- •Awareness sessions
- •Statement of union-management cooperation
- •Employment security assurances
- •Cross-functional task teams
- •Joint union-management vision statement
- •Pay for knowledge
- •Group-process training
- •Group-process joint union-management steering committee

were only reinforced during difficult collective bargaining over job flexibility. No matter what was taught or communicated about the Champion Way or Participative Management, it was greeted with suspicion and cynicism. The performance of a once-proud mill was deteriorating, and the start-up of their critically important #4 paper machine was floundering after a major rebuild. People seemed to want to return to a livable equilibrium—as one employee put it, "It's really hard on you not to want to do a good job"—but no one was willing to bridge the gap. A wake-up call came in 1990 however, when representatives of a major customer, *Time* magazine, came to the mill and met with leaders of both management and the union, citing statistics unfavorably comparing Bucksport quality with that of the competition and threatening to take away their business. Ralph Burgess, then president of the union local, reported that as a result of that meeting he and management realized that the survival of the mill and his members was at stake; they had to find a better way of doing business together.

That realization led to a whole range of changes inside the mill, including training and redesigning of work structures; but most important was an ongoing effort to keep employees in touch with customer needs, including direct contact through crew visits to customer locations. "Customer partner teams" and "supplier partner teams" were formed. The "union-management leadership team" began operating and increasingly developed a relationship built on trust and a common stake in a better future. Bucksport was the first major mill to employ what came to be known as "interest-based problem solving" in reaching agreement on their 1990 contract (see Chapter Three for a discussion). Within the company, the Bucksport mill became a leading example of a successful change effort in the early 1990s.

Another company benchmark in Customer Driven Quality was Champion's oldest mill in Hamilton, Ohio, where management of both sales and operations combined to lead an effort they first called "Market Talk," where the emphasis was on direct communications between Champion employees and customers, and, later, "Market Talk/Market Action," where customer partner teams began joint problem solving. One manager commented, "I wouldn't have believed the progress we've made with Market Talk/Market Action and partnering"; while another said, "The people involved in Market Talk really know and understand our customers and how important that understanding is to our success; our goal is to get a lot more people involved." An hourly employee told of how "on my Market Talk trip we saw how our customer was planning to substitute acrylic for some of the things he used to use paper for. We have to find new uses for our paper, or other materials will drive us out of business." The power of this straight-from-the-horse's-mouth communication in creating alignment was brought home by a clerical employee who referred to earlier change initiatives as "team programs" that were "management driven" and noted that "people did not necessarily believe management about the problems." However, he said, "Market Talk is customer driven, and you had better pay attention to what the customer says."

These illustrations reflect a larger change in context for the whole company. Champion held its annual management meeting for its top managers in October each year. In the fall of 1989, a sales manager who was not at the meeting said that he heard that the word *customer* was not used once in the three days of presentations and discussions. The next year (and succeeding years) the meeting was expanded to include more operating, sales, and marketing managers from the field, including the manager who made the comment above, and the word *customer* was probably used more than any other, as the focus was on customer-driven quality and continuous improvement.

Planting New Ideas in Off-Site Sessions

Hundreds of thousands of person hours were invested in discussing the objectives and methods of the change process in many different types of forums in many places over the decade. One such forum was the management-development program, the "GTE experience." It may seem strange that a mechanism for alignment in one company would use the name of another company. The name came about when Champion developed a comprehensive management-development curriculum in the 1980s and needed a place to "house" it. GTE owned a complete residential conference center in Norwalk, Connecticut, but was in the process of transferring its headquarters and much of its staff to Dallas and, while continuing to use the center, would no longer need it full time. Champion entered into a licensing arrangement and now uses the facility extensively, with all managers required to "go to GTE" to progress through the management-development curriculum.

The substance of this curriculum and its impact on change will be a subject for discussion on capability in Chapter Three; however, we include it here because of the powerful effect the GTE experience itself has on achieving alignment of managers across the company. When they enter the retreat setting for a week or two, participants leave their position and location hats at home to some extent and put on the common hat of a Champion learner. They become exposed to common models and engage in cross-location problem-solving exercises and other projects. The top executives stop in for a dinner with each group, after which they engage in an informal question-and-answer session. Coffee breaks, meals, basketball games in the gym, and late-night get-togethers are all vehicles for creating a sense of being part of one company. All in all the experience serves Champion in the same way the better known "Crotonville experience" serves General Electric.

Aligning Structures and Systems

From the beginning of the change process and central throughout were revisions in organizational structure and the design of work to create alignment with the Champion Way statements as well as with business goals and strategy and the technology itself. These realigned structures were in turn reinforced by innovative changes in pay systems, communication and information systems, and planning systems.

Creating Work Structures That Fit

Over the decade Champion's work structures were fundamentally redesigned at every level and in virtually every functional area. The changes are explored here for their alignment effects—that is, their fit with Champion's competitive strategy, core technologies, and social values. How some of these changes were formulated and implemented through participative processes will be explored in Chapter Four.

The first structural change took place early on, in 1986, when the corporate structure went from a centralized functional form to business units organized by broad market areas such as printing and writing papers, publication papers, newsprint and kraft, and forest products.

At the workforce level, other mills followed the lead of Quinnesec and Pensacola in using the process of sociotechnical (re)design to devise "whole" jobs and work systems tied to the needs of internal and external customers. In the course of the analyses performed by design teams, the core technology was carefully examined in the light of business strategy and the values of the organization. The analysis of process variances in this approach helps, at the redesign stage, to align individuals' jobs and mind-sets with work-unit technology, increasing the likelihood of effective cooperation within the unit. Similarly, alignment of groups with a mill's core technology promotes understanding of the interdependence of these groups.

This process of redesign didn't take place only in mills however. Early in the change process, the corporate control department (accounting) redesigned its more than twenty specialized functions—such as payroll, accounts receivable and payable, and cost analysis—into five multifunctional customer teams (one for each of the new business units and one corporate), thus aligning the accounting infrastructure with the structure and strategy of the business. They changed their compensation and career-path systems to align with this more horizontal and multifunctional team approach, and, as the department matured, they moved into peer performance evaluation and self-directed work teams.

Another early adopter was corporate technology—the central R&D group—and alignment with business purpose was at the heart of their redesign. At the beginning of their process, they identified "lack of business perspective" in their technically oriented culture as one of the key barriers to their success. They developed a business- and customer-oriented vision by consensus; flattened the structure (from five layers to two) with self-managed teams focused on internal customers—the business units and mills—through the newly created role of business-unit coordinator; provided general business education and communication; and again made significant changes in the reward and career-management systems. The net result was that strategic alignment with internal customers became a strength of this organization. According to Gerard Closset, the manager who led this change, and Marlene Feigenbaum, the internal consultant,

> Over the past five years, our internal customers have increasingly solicited our support, involvement and direction in both the traditional problem-solving role as well as the more strategic developmental role. This expanding role has been accompanied by a gradual growth in funding support. Our internal customers have described the improvement of the quality and delivery of our services as follows:
>
> • Better focus on projects requiring unique expertise
>
> • Innovative approaches with direct impact on competitiveness and profitability

- Faster, more effective response to field problems

- Earlier and more accurate identification of competitive threats and opportunities

- Greater competence, commitment and energy across the organization[1]

They continue, "These are the primary reasons why most technology people no longer feel on the sidelines—why they are developing a strong sense of belonging to the whole of Champion."[2]

This sort of business focus expanded to other technical areas in 1994, when the entire set of functions reporting to Dick Olson, then executive vice president of all the technical staffs, were redesigned. The corporate technology unit described above along with corporate engineering and manufacturing support were combined into the "applied technologies organization," a flattened structure oriented to the business units with cross-discipline teams focused on either strategic or business-maintenance projects.

This realignment of support infrastructure around business units took place over several years, although there was a moment of doubt. In 1988 the mills were not running well at all; the senior executives debated whether the business-unit structure was part of the reason and considered returning to the centralized functional structure. The business units were only a couple years old at the time, and the support infrastructure was just beginning to make the kind of shifts we have been describing. A time lag of this kind is normal and predictable but often is not allowed for as organizations constantly change direction at every alarm. Fortunately, in this case, the decision was made to continue with the business-unit structure, and the alignment did eventually take place. The business-unit structure remained until late 1996, when, as discussed in the Prologue, the Gang of Eight created the new matrix of functions and business teams.

As the realignment of infrastructure matured to support the business units, these units and the mills that constituted them became increasingly aligned with their external customers through a variety of arrangements. For example, middle management of some mills moved from departmental functional organizations to interfunctional "grade-line" or "business-systems" structures focused on specific products and customers. Such structures are often called "focused factories" in the literature on organizational design. As another example, the business units developed cross-mill "business teams" organized around market segments; some of these teams developed customer partnerships. The cross-functional membership of these business teams included several managers from operations, but the teams were led by sales and marketing to guard the customer focus. The experience with these ad hoc business teams led the Gang of Eight to formalize this sort of structure in 1996.

Redesign at the senior-executive level of the corporation began fairly late, about seven years into the change process (1993–1994). This timing is consistent with the findings of Michael Beer and his colleagues. After presenting evidence that most genuine organizational renewal does not begin at the top and cascade down, as the conventional wisdom would have it, but rather begins with innovation in the field, they ask

> But what about the top management team itself? How important is it for the CEO and his or her direct reports to practice what they preach? . . . Senior managers can create a climate for grass-roots change without paying much attention to how they themselves operate and manage. And unit managers will tolerate this inconsistency so long as they can freely make changes in their own units in order to compete more effectively. There comes a point, however, when addressing the inconsistencies becomes crucial. . . . This last step in the process of corporate renewal is probably the most important. . . . The time to tackle [this] tough challenge . . . comes finally at the *end of the change process* [italics added]. At this point, senior managers must make an effort to adopt the team behavior, attitudes, and skills that they have demanded of others in earlier phases of the change. Their struggle with behavior change will help sustain corporate renewal in three ways. It will promote the attitudes and behavior needed to coordinate diverse activities in the company; it will lend credibility to top management's continued espousal of change; and it will help the CEO identify and develop a successor who is capable of learning the new behaviors. Only such a manager can lead a corporation that can renew itself continually as competitive forces change.[3]

When the Champion senior executives first met in January 1993 to begin their process of self-examination and redesign, Sigler in essence made the same points as those above, asserting that "by now it's clear that teams work at other levels; why can't we try it at ours?" He added that he, President Whitey Heist, and several others at the senior staff level would be retiring in the next few years, and they needed to prepare for leadership in the future. This process eventually led to the formation of strategic teams in the top few layers of the company. After these teams had been working for about a year, Sigler acknowledged the power of team management and noted that the change process had redefined senior management's role too. "We've worked hard to understand how teams manage, and we think we do understand organizational structures better," he said. Rather than being tied into a hierarchical, controlled system, Sigler pointed out, senior managers now address problems that fall between the cracks "because of our unique perspective of seeing and understanding how the total company works together."

None of these design changes were in any sense quick fixes. Rather each redesign occurred as the units in question became ready to take on the task. Moreover, the beneficial effects occurred only as people developed new mind-sets and related to each other and their customers in new ways. The common thread reflected in all these changes was a more systemic and less bureaucratic mental model, aimed at optimization of the whole rather than maximization of the parts. As this thinking matured, a new phrase was coined: "we must move from being an organization that is functionally efficient to being one that is business effective."

Reinforcing Alignment Through Rewards

What people get paid and rewarded for sends important signals about what is important to the organization and therefore can be a powerful force for alignment or misalignment. In Champion compensation became another area for reexamination and change. Pay for skills and contribution was implemented in conjunction with work redesigns to facilitate multiskilling. Some of the criteria by which each mill was judged to determine management bonuses reflected priorities in the change process. A corporate task force did a thorough study of gainsharing and recommended guidelines that were adopted by the transition team and communicated to the locations. Each mill could then decide when and whether gainsharing was appropriate for them and could develop and implement their own plan with corporate review and approval. The first such plan was implemented in the Alabama forest-resources unit, followed by the Quinnesec mill. The first plan developed and approved on a joint union-management basis was at the mill in Sartell, Minnesota, in the fall of 1995. The alignment purpose is highlighted in the Sartell plan's statement of objectives: "gainsharing should provide employees with a clear picture of the tie between what they do, how the location performs, and how they are rewarded."

In 1995, as the industry emerged from a deep trough, Champion and the UPIU announced a one-time incentive plan to encourage a performance breakthrough. Called Something Extra, it rewarded all employees at each location for meeting organization-wide production goals. Something Extra goals were set high, above the officially sanctioned targeting plans, and yet were so consistently exceeded that additional (Improvement Plus) incentives were offered and paid. (Gainsharing, Something Extra, and other reward systems are covered in depth in the next chapter.)

Reward systems other than compensation were also affected. For example, in functions such as R&D and corporate accounting, career paths were clarified and were based more on multiskilling and professional development than on hierarchical promotion. In fact the entire "pay and performance management" system was studied and made more flexible and relevant to organizational goals. Managers were promoted, demoted, terminated, and transferred based in part on whether they performed in a way consistent with the values and principles of the change process. Promotions and transfers provided opportunities to move people who had positive experience with the new model into units where change leadership was needed.

Information for Order and Alignment

From Mary Parker Follet's concept of the "Law of the Situation" in the 1920s to Margaret Wheatly's application of "complexity science" to organizations in the 1990s the importance of information as a force for order and alignment in organizations has been stressed.[4] The work of Kathy Dannenmiller and her colleagues on "large-scale" change emphasizes that effective strategic alignment is possible as long as everyone is operating from a common database. However they point out that it doesn't take long in today's world for the "commonness" of that database to erode, therefore the on-going need for efforts to maintain it.[5] All the initiatives described above carried a large commitment to communications, from the one-on-one interpersonal level to meetings large and small to the use of print and other media. Open-books management became the practice at Champion. All business information was openly shared, and education was provided on how to understand and use it. Concepts such as free cash flow and debt-to-equity ratios were explained to all levels along with the relevant numbers for Champion and, as appropriate, its competitors.

Information systems were adapted to fit the evolving organizational designs; for example, instead of functional departmental reports at the Pensacola mill, monthly profit and loss statements were organized by "business system." A companywide cross-functional team performed a thorough study of how information systems could be redesigned and used to (1) run the mills better, (2) enhance net revenue, (3) reduce overhead, and (4) reduce interest cost, in that order. They asked questions such as Do the data exist to accomplish these goals, can they be captured, and how can they best be transmitted to those who need them? With regard to the last question, the emphasis was put on the horizontal flow of information rather than on the more typical vertical, hierarchical flow.

In addition to information systems, general communications were a multimedia undertaking. Newsletters were used at every mill, sometimes jointly sponsored and managed with the local unions. At the corporate level, in addition to the companywide newsletter *The Wire,* a videotape called *Inside Champion* was mailed monthly to each employee's home.

Targeting to Raise Aspirations and Set Plans

A new planning system called "targeting" was introduced in 1986, the same year that diffusion of the change process began in earnest, and they came to be closely tied. Targeting was Champion's approach to strategic and tactical planning by unit. Each location and corporate department was required to develop performance targets and plans for achieving them on a rolling five-year basis. The first year targeting was done by the field units in a fairly perfunctory way as a "number-crunching exercise"; the plans were put together largely by the location controllers to satisfy the requirement of headquarters in Stamford. By the second year, however, the leaders of the Pensacola mill decided to make targeting a vital part of their change process; they used it to explore values and philosophy as well to develop a "strategic business model" (a specific operational scheme for how the mill would look and run in the desired future state) and to identify key performance indicators (KPIs) as a foundation on which to plan. This process mostly involved top mill management in 1988, with some input from other levels, but in subsequent years targeting increasingly became a vehicle for employee involvement at all levels. A similar targeting approach gradually spread to other locations over the next few years and eventually became the norm. It served as a powerful alignment tool for integrating the many initiatives in a systemic way not only at the time of the annual planning cycle but throughout the year, as targets in the form of KPIs were now monitored.

The Sheldon, Texas, mill took the alignment power of targeting to a new level in 1995. First the senior management group and the executive board of the local union went off-site for a participative process called a search conference, as developed by Eric Trist and Fred and Merrelyn Emery. These meetings involve formulating an "active-adaptive" plan for the future that takes into account the turbulent environment in which business currently takes place. Six major targeting goals were jointly developed in this meeting. The next step consisted of two large-group two-day conferences in which over three hundred employees actively participated in "real-time strategic planning."[6] These "deep-slice conferences" included employees from every level and function in the mill as well as from units of the company that supply raw materials and services, business-unit and corporate managers, and representatives of the international union. After the plans that grew out of these first two steps were reviewed by the corporate executive group, every Sheldon employee was involved through further large-group interactive sessions in translating the millwide targets into crew-level objectives and plans.

Aligning Links Across the Value Chain: Courtland #35

An impressive feature of Champion's process was the fact that it affected every link in the value chain from vendors to various groups of managers and workers to customers. We will underscore the power of alignment by presenting a case that demonstrates alignment across the value chain, concluding with a discussion of key insights that apply not only to the particulars of this example but to the larger Champion story.

The design and start up of the #35 paper machine at the Courtland, Alabama, mill, the largest (and therefore the flagship) of Champion's eleven domestic mills, represented, in the autumn of 1993, the culmination of developments and lessons learned in the change process to that point—a process that encompassed both the massive capital program and the change in organizational culture. To some extent, the whole history of change at Champion crystallized in the start-up of the new 240,000-ton-per-year machine.

Involving Engineering and Construction

By the time the board of directors approved funding for Courtland's $400 million project in November 1990, experience with several other major projects in the capital program had taught the engineers and operating managers the importance of attending to the human side of planning new facilities. On some of the early projects, they had not paid sufficient attention to the human side and had suffered the consequences; experience with later projects had shown them that effective involvement of people paid off.

Among the lessons learned was not only the value of building effective project teams but also the value of beginning that process early on so that, at best, problems could be prevented from developing or, at least, could be caught before they required major intervention. Another lesson was the importance of a clear project scope, which, for #35, was the high-speed machine and its necessary support systems. Consistent with sociotechnical principles, the machine design would provide for self-regulation of the work crews by integrating the winder operations—work and equipment that were traditionally located in another department.

The engineering phase of the project brought together a talented group of individuals from Champion engineering, Courtland mill operations, and the consulting engineering firm of Brown & Root. All had been involved in previous successful paper-machine projects; thus #35 offered an opportunity to build appreciatively on past and current successes and bridge from them to an even more desirable future, based on the assumption that these bright, skilled, and experienced people already knew how a good project organization worked.

The team included not only Champion and Brown & Root engineering and construction people but also representatives from the Courtland mill operations and major equipment vendors, such as Beloit, the manufacturer of the paper machine. In an initial team-building workshop with the top forty people, they narrowed the factors and conditions for success down to four: teamwork, leadership, trust, and ownership of common objectives. From that point on, the whole project—team-building events, training in team skills, the development of a project mission and philosophy statement, the development of project measurements, a mid-point project assessment, and even the everyday management and atmosphere—reflected a commitment to and a confident belief in the ability of this team to deliver a world-class project. This confidence is illustrated in these comments from interviews done as part of the mid-point assessment:

- Maybe we have, on this project, the cream of the crop. I remember projects where each discipline was sacred, but not this one.
- [As a result of] the team-building experience we all pulled together. I've never been on any athletic team that came together like we did—total communication and cooperation across the board.
- There is a positive attitude on this project from the top down.
- Communication on the #35 project is the best that has ever been at Brown & Root.
- Communication is emphasized, constant feedback.
- The time given to engineering to be ahead of construction is making the job run smoother and is allowing for better quality.

Team building continued into the construction phase, and, about six months before the machine was scheduled to start up, a "future search conference" was held; it included over sixty people from all the different levels and parts of the project: vendors, mill mechanics, engineers, construction employees, mill managers. It began with a look backward at recent history, starting from a global perspective and then gradually zeroing in on the paper industry, Champion, and finally a personal perspective. From these data participants projected trends and developed two visions for the future that encompassed both the technical and the social systems: a short-term vision of the elements of a world-class machine start-up and a longer-term vision of that same machine operating successfully well into the future. The last step in the conference was the redesign of the project organization from "discipline teams" (such as electricians, millwrights, and instrumentation specialists) into "system check-out" teams (such as for the calendar pulper or the rewinder) for the final pre-start-up phase of the project.

Empowering a Participative Organizational-Design Team

Paralleling these developments in engineering and construction, another team was actively engaged in getting ready for #35. The "core group" consisted of twelve Courtland employees chartered to design the work organization and eventually become part of the machine crew. Eight of the twelve were hourly operators or maintenance employees, and the rest were middle managers and supervisors. Starting their work in July 1992, over a year before the machine was to start up, they visited several high-performance mills outside of Champion, were briefed on plans for #35, and studied the technical and social requirements for the new work system, comparing them with those they were used to on existing machines.

The core team developed the Five Cs philosophy (see box), a statement of social values consistent with the Champion Way; based on these principles they designed an organization that would, according to a process coordinator (one of the new roles), "control the entire system, from stock preparation to roll wrap to meeting the customer." That degree of self-regulation had been made possible by the innovative technical design described earlier, which

The Five Cs: 35's Philosophy

CUSTOMER FOCUS: We partner with suppliers and customers to ensure our place as industry leaders.

CONTINUOUS IMPROVEMENT: We constantly improve by listening to new ideas and sharing our knowledge through ongoing training and teamwork while maintaining a focus on safety, quality, and production.

COMMITMENT: We are an organization of goal-oriented employees who are encouraged and empowered and strive to do things right the first time.

COMMUNICATION: We understand the value of effective communication, are not limited by traditional barriers, and talk openly and honestly at all levels.

CONTROL OF THE PROCESS: We recognized the value of product consistency and use statistical methods to control operating variables. We are empowered to make the decisions necessary to reduce product variability. We expect high quality from our suppliers and anticipate our customers will expect the same from us.

integrated roll wrap with the machine, a good example of joint optimization of the technical and social systems.

Over six hundred employees in the mill changed jobs as a result of this expansion, creating an immense need for training. At the same time, the training provided an opportunity to rebuild pride in the Courtland organization. All members of the crews for #35 were trained in communications, problem solving, and team skills in addition to receiving in-depth technical training. They also visited outside suppliers and customers. With internal suppliers and customers, such as the maintenance department, shipping and warehousing, and production control, they clarified expectations about how they would work together through written agreements.

Consistent with sociotechnical principles, the new design called for multiskilling, with cross-training built in and rewarded with a pay system based on skill and contribution. As the rewind coordinator said at the time of start-up, "With overlapping job responsibilities, traditional boundaries are gone. If you see something that needs doing, you can step in and help. Everything moves forward."

Also layered into the compensation scheme was a form of gain-sharing that would kick in as start-up performance goals were met.

Partnering with Customers

Five of the largest customers were asked to discuss their needs with Champion before the machine came on line. These customers also made plans to conduct trials at their plants using paper from #35. A workshop was held for mill and project representatives together with personnel from sales, marketing, and field services to prepare for taking the product to the marketplace and to gain agreement on how customers would be served. All the #35 operators visited customers to see how the paper was used in their plants in order to understand what customers needed and expected. "I've seen many new machines start up," said the director of purchasing for one customer. "What's unusual in this case is Champion's dedication to being at our plants for trials to see how their paper performs and provide instantaneous feedback to the mill to make any necessary corrections." Customer partnering went on not only on the product level but in terms of philosophy as well. As the national purchasing agent for another major customer said, "Their team concept is one that our own plants are moving toward, with the focus on the customer." His company helped qualify the #35 machine by testing its paper at its own customers' locations.

An Impressive Start-Up

The mood throughout the project was upbeat and positive, but the proof would be in whether the machine started up and ran successfully. Courtland #35 presented the opportunity to integrate all that had been learned from earlier projects and to "do it right."

Based on some of what had been learned, project management resisted setting an arbitrary start-up date. The executive vice president in charge of the project, future CEO Dick Olson, made it clear that the important goal was a machine that would run well for years into the future. "We're thinking thirty years down the road," he said, and that goal should not be compromised to meet some start-up date arbitrarily arrived at months or years before. However, sales needed to let customers know when they could start expecting paper, so a general start-up window was established for late August 1994. Bill Burchfield, executive vice president of printing and writing papers, the business unit to which the Courtland mill reported, had to balance these needs, but he clearly supported the central message: "start it up right."

Relieved somewhat of external pressure to meet a deadline, members of the project team nevertheless felt their own internal pressures to meet their goals regarding a first-class start-up of a world-class machine. After weeks spent carefully testing out and debugging systems, finally the day arrived to begin the production start-up. The news was covered in a special report in *The Wire*. "In the early hours of Sunday, August 29, 1993, a new chapter in the history of Champion International Corporation was written, with the successful start-up of the company's thirteenth uncoated paper machine. The No. 35 machine, at the Courtland, Alabama, printing and writing papers mill, roared to life at 4:01 A.M. The start-up culminated Champion's two-year, $395 million investment at Courtland, which is now the largest fine paper mill in the world."

Results of the start-up were more than impressive. Forecast tonnage at start-up was 492.2 tons per day. The ultimate target was to make 703.6 tons per day by 1997. Actual performance exceeded plan every month, with some days over 800 in the first month of operation. And all grade lines quickly met qualification tests. Jim Cooper, the paper-production manager, summed up a lot of the feelings in the mill when he said, "The machine is running well, and quality is excellent. After two years [of getting ready], we're all lightheaded with excitement and pride." And the vision of long-term success on this machine has been fulfilled. As of this writing, performance continues to be world class.

Putting It All Together

"On #35," said Bill Bannan, vice president and Courtland operations manager through the expansion period, "we had a unique opportunity to demonstrate what can be achieved when you invest as much in people as in engineering. . . . We've taken employee empowerment and customer partnering to a new level. . . . Union and management at the Courtland mill are working together and moving forward."

Much of what Champion had learned to this point in its total systems approach to change thus came together in one place at one time. The great investment in the notion of joint optimization was realized through the simultaneous and integrated design or redesign of both the core technology and the social systems. Improved process control was demonstrated in the engineering and construction processes, the social-system design process, and the order-to-delivery chain, including papermaking. Customer focus was at the center of it all. Moreover, compatible with all those desired outcomes were the means by which they were accomplished: a process of empowerment that encompassed customers and suppliers, executives and managers at all levels, union leaders, engineers, construction specialists, production and maintenance employees, and administrative support staffs. A variety of methods were used, from team building and work redesign to appreciative inquiry and search conferences; but all were consistent with the integration of the core technology, values, and business strategy in a way that reached across the value chain and into the customer's place of business.

Summary and Conclusions

We have discussed a large array of devices Champion used to align its change process with its competitive strategy, its core technologies, and its social values and beliefs—the alignment touchstones. Some devices were intended primarily to align mind-sets or attitudes—language, conceptual models, and training and development sessions—and others were aimed at behavior—structures, rewards, information systems, and planning systems—although in the end all these devices were assumed to influence both attitudes and behavior. Devices such as the reward and planning systems were aligned with each other by a top management team. Eventually, as illustrated by the #35 machine at Courtland, many other techniques, such as search conferences and the work of numerous groups, were combined in real time to produce an extraordinarily successful start-up of a strategic capital investment.

Based on our observations of Champion and other corporate change efforts, we conclude that the key to sustained and comprehensive change is aligning the organization and change processes to all three touchstones (each of which involves high aspirations) and doing this through many redundant devices, while ensuring that these devices directly involve all parts of the organization and beyond to vendors and customers.

This chapter sets the stage for our discussion of capabilities and letting go in the chapters that follow. Management spent much time and effort on alignment because if individual initiatives are not aligned—are not aimed in the right direction—then stronger capabilities will be wasted and greater latitude will lead to confusion. Stated positively, clear alignment both permits greater latitude and ensures that capabilities will be put to strategic use. We turn now to how organizational capabilities were built at Champion.

Chapter Three

Building Capabilities
Competence, Commitment, and Cooperation

*She comes home at night with books to study about her job.
It's like she's getting a junior college education out here as
she prepares herself for advancement and to double her pay
from her entry rate.*
—Mill supervisor speaking of his daughter, who was
hired in a redesigned work area, November 1991

*What I want to stress is the culture that has developed here
because I think that's really what it is all about. I've never
seen a culture anywhere else where there is so much
mutual respect—level to level, area to area, and person to
person. Everyone takes a total system perspective.*
—Forester, August 1992

*We can either work together to straighten things out or put
a padlock on the door.*
—Local union officer, June 1992

Because the stories of individuals sometimes tell the stories of companies vividly and concretely, we begin this chapter with the story of Dennis Didier, a manager in the Pensacola mill. His story illustrates our entire "what works" framework. Woven throughout the case are elements of alignment, of capability, and of letting go, as well as of their interactive effects. Although we focus in this chapter on capability, we must stay mindful of this larger context (see

Figure 3.1). However, the Didier story emphasizes how his development included the Three Cs of Capability—competence, commitment, and cooperation—which are the subjects of this chapter. This story also illustrates our continuing theme of a systemic approach to change. Immediately following the Didier story, we amplify this theme by discussing a major mental shift that occurred in the way managers and other employees approached their work—a shift toward "systems thinking." In the remainder of the chapter, we describe and illustrate some of the many initiatives taken at Champion that increased capability, including training and education, rewards and recognition (including employment security), and union-management relations.

Case History: The Transformation of a Manager

Dennis Didier was a process engineer by training, experience, and inclination. He was involved in coordinating several capital projects in the Pensacola mill, first with St. Regis and, then, after the merger, with Champion. The projects included, in 1988, the rebuild of P3, an older paper machine that was converted from making stock for paper cups to making fine-paper products such as off-set and envelope papers—products that were more in line with Champion's business strategy. As that project was coming on line in 1989, he accepted the position of P3 machine manager and soon (1990) became complex manager of both P3 and a rebuilt pulp dryer.

As he moved from purely technical roles to operations management, he carried with him his technical view of the world and

Figure 3.1. The Performance Formula: Capability.

Alignment ×	**Capability**	× Letting Go = What Works
	• Competence	
	• Commitment	
	• Cooperation	

ran his area as a manufacturing function focused entirely on keeping the equipment producing tons of product. He "first got religion," as he puts it, when he went to GTE (Chapter Two) and participated in a two-week course in the management-development curriculum along with other managers from a variety of functions from all over the company. "For two weeks I sat there," he said, "and listened to all these people from marketing and human resources discussing how we should be thinking about the business and customers. Our instructors from different universities were pushing business models and market strategies. All of a sudden it just hit home, and I went from thinking 'I make paper' to 'We have a market out there.' At that point I didn't know and couldn't have named one of our customers.

"When I came back to the mill, I went out to P3 and asked my foreman what customer we were running paper for that day, and he said, 'Champion.' I knew that wasn't good enough anymore, and he and I went through the order book and looked up who the real customer was. Up until then, about the only time we heard a customer's name was when we had a complaint.

"Then lots of other things happened. In 1990 I became part of a targeting team to look at the future of P3. Obviously we couldn't compete with the big machines like P5, and we couldn't make it just pulling pulp stock on the dryer. So we began to think maybe we should be in other grades of paper than just off-set and envelope, but sales told us no. Not too long after that Kim Dean became a product manager for sales and marketing, and she came to us with a request from a customer for electronic-data-processing label paper and asked, 'Can you make this stuff?' We developed it with the customer in four or five months. A little later we did the same thing with return post card."

By that time, Pensacola mill management as a whole—through its experience with redesigns at the workforce level, Customer Driven Quality, millwide targeting, and managing by trends in KPIs—had shifted from thinking functionally toward whole-system awareness. Without yet changing the formal organizational structure, they began to think in terms of four "production systems" by product line: pulp, cut-size paper, off-set paper, and market pulp. Thus, when the targeting cycle began again in 1991 to plan for the five years from 1992 to 1996, Dennis led what was called the P3 System Targeting Team. He recalls, "We had members from different departments asking questions like what grades should we be in and what are we good at. We were still pretty functional, but this time, by working *with* sales and marketing and their knowledge of customers, it all started to make sense.

About then a new mill manager, Doug Owenby, was transferred from Quinnesec, Champion's greenfield team-based mill in Upper Michigan. Although he strongly emphasized principles such as team work and involvement along with high personal and business values, his operating management style was fairly hands-off, providing considerable latitude for middle managers to run their own show. According to Dennis, "Doug creates a great environment to work on this stuff—enough freedom to operate. Suddenly I felt really empowered. For example, we can take lost time on a paper machine while we're working on development of a new grade. A pure papermaking manager wouldn't let you do that. So I could run things the way I thought best. I had developed good relations with the salespeople, and even though we still had a functional structure, we started operating informally as a system."

Reflecting this evolution in thinking, in 1994 Pensacola mill management formally redesigned all management and support-staff functions, which resulted in almost all salaried employees' being assigned cross-functionally on a full-time basis to one of the four systems. A relatively small number were left in central, mill-wide functional groups. With the broadening of conceptual boundaries to include order services, sales, and marketing, and with an increased focus on markets and customers, the four areas were no longer called production systems but, rather, *business* systems. Dennis was officially named "off-set system leader." "Today," he says, "almost all functionalism is gone. Everybody knows how they fit in. Almost everybody knows all the customers. We're not out here just making paper but serving customers through an order-delivery chain. A quality product is almost a given; we're in the service business now. The customer is part of that same chain. Most of our customers are not end users; they have customers that they have to supply. We now understand that whole chain and our place and their place in it. We used to only talk tons; we now talk about order adherence—quality, quantity, and delivery. It drives your business in a different way. We used to only care about our product until it went off the end of the machine. Now common focus and understanding eliminates all the turf battles. On quality, everyone knows what the customer needs, and if we're going to err, we all know to err on the side of the customer. We have become much more responsive. We used to change grades about once every eighteen months. Now we're tinkering with grades all the time. We used to make vanilla and chocolate. We still don't have thirty-one flavors, but we can make you chocolate with nuts or whipped cream on it."

Through such activities as his GTE educational experience, his contacts with sales and customers, his work across functions in the mill, and his goal-setting and planning experience in the targeting process, Dennis increased his knowledge and skill *(competence)*. All this participation and involvement took place within the framework of a changing organizational design, climate, and culture at the mill, including the arrival of a new mill manager who provided room for Dennis and his team to operate semi-autonomously, thus increasing Dennis's sense of ownership and *commitment*. The change toward a holistic, systemic view required new levels of *cooperation* within his own cross-functional team, with mill management, with functions outside the mill such as sales and order services, with other mills, and with customers. We think of competence, commitment, and cooperation as the Three Cs of Capability, and they are the subject of this chapter. As our framework suggests, alignment with purpose, although critical, is insufficient without skilled, dedicated people working together to realize that purpose.

The Didier story also adds continuing emphasis to our theme of a systemic approach to change. Not only did he transform his thinking, the design of his organization, and his behavior from a functional to a systems orientation, but the process of change itself was systemic. Organizations are often frustrated with the ineffectiveness of what appear to be sound programs such as management training. They send managers to good programs to teach them new concepts and skills, and then, when the managers return to business as usual, they wonder why the programs didn't take. However, Dennis, as a target of planned change at Champion, was immersed in many complex interventions: the GTE management-development program, targeting, organizational redesign, Customer Driven Quality, a new boss who managed boundaries rather than micromanaged, and others. These were not the random flurry of unconnected activities that characterize improvement or transformation efforts in many organizations and that so often wither and die. They were multiple, mutually reinforcing efforts directed toward the same goal, and they produced the critical mass necessary to overcome the natural inertia present in any organization. In this case, the goal we are addressing is change in a personal system by the name of Dennis Didier, but the same holds true of organizational systems.

Appreciating the Whole: From Functional Efficiency to Business Effectiveness

Successful experience with systems designs on the shop floor influenced the move toward systems design at the middle-management level of the mills. In the past the norm in a mill would have been for the manager of the power and utilities department to go to the regular morning production meeting ready to report on his area's performance and to attend to the reports of others only when they might affect his area. All he cared about was that his area performed well; if it didn't, he might be embarrassed or receive criticism from his boss or jibes from his peers. Over time these morning meetings began to include discussions of budgets or capital plans; the manager of the power and utilities department was now asked for his input on spending in all the other areas of the mill, and his peers had input on his spending. As they got further into targeting, he was asked to help shape the values, vision, and long-term goals of the mill as a whole through consensus with the mill senior-management team. That process eventually began to include related functions beyond the mill: timberlands was invited to participate, and then corporate scheduling and sales, and maybe R&D. Local union leaders joined in targeting, and they, too, began to see the "big picture." Increasingly, they were all focused on the customer and shared an understanding of how all the parts needed to work together to deliver what the customer wanted and to do it profitably.

Just as success on the shop floor influenced these developments in the middle, the success experienced by middle managers
like Dennis or our typical manager in the earlier story influenced
the business-unit level. Joe Donald, then executive vice president
of the publication papers business unit, was committed to moving
Champion from "a functionally efficient to a business-effective
organization" and saw the business-system model as a way to do it.
In early 1993 he sponsored the formation of a business team
focused on the coated-papers product line. Bill Burchfield adopted
this approach with business teams in his business unit as well; this
was the unit to which Didier reported. These teams were not only
cross-functional but also cross-mill. In addition to his role as system leader at the Pensacola mill, for example, Didier became a
member of the business unit's off-set core business team.

"In the past," Dennis reports, "the product managers from sales
and marketing would have to make the call on blending manufacturing capability with customer requests all on their own, with
the operating people trying to buffalo them about what could and
couldn't be done. Now we have a cross-functional, cross-mill team
trying to make that match in a way that makes the most money for
Champion International Corporation as a whole. We manufacturing types now understand a whole lot more about what salespeople face. In the last two years I've gotten to know 90 percent of the
district managers and have visited a lot of customers. And because
I am part of a systems team back at the mill, I can meet with the
core team and go back and make something happen. It can create
an exciting sense of achievement for everybody. Our core team has
a big sense of entrepreneurship—running a small business inside
of a big company."

In turn, the senior business-unit teams began to tackle issues
and goals at their own level in this systemic way. We already
described in Chapter Two how corporate staff groups began to
realign themselves in support of the business units and their focus
on customers. There was a growing horizontal as opposed to vertical mind-set and cooperation increased across the value chain.
Finally it began to happen at the top, with the on-going series of
Treetops meetings of the senior-executive working group that
began in January 1993. There the top managers of functional
departments and business units began to look at the total strategy

of the company and not just their own areas. As described in Chapter Two, they sponsored strategic teams with membership from across the company. At the annual management meeting in October 1995, business-unit and staff executives made presentations on strategic imperatives not, as they would traditionally have done, for their respective units but for the corporation as a whole.

The effect this evolution in the boundaries of cooperation had on performance would be impossible to calculate, and, in any case, much of it remains to be played out. We would hypothesize that its impact has already been profound and will be more so in the future. Not to be underestimated is the powerful effect this evolution has had on the growth of the individuals involved. New mill managers once were selected from among candidates who had experience in one or two functions; now the pool consists of candidates with a broad business perspective. Union officials can provide better leadership for their constituents knowing how the total business works and having total business information. Managers from staffs and locations who have been involved in corporate-sponsored strategic teams gain knowledge in areas to which they never before would have been exposed. Finally, those who were replacing the six executives at the top who would retire in 1996, including the CEO and COO, had the opportunity to develop an understanding of the whole business and hone their skills in shaping a large, complex enterprise for an uncertain future. Their experience with business teams led them to institutionalize similar structures when they took over as the Gang of Eight in the fall of 1996. Through the growth of systems thinking, then, people gained new competencies, became increasingly committed to the total business success of the whole organization, and cooperated with others across boundaries heretofore impenetrable.

Upgrading Competencies and Aspirations: Training

According to the Champion Way in Action statement, "To create the working environment we seek, we will . . . train and retrain employees at every location to be certain that everybody is qualified to fulfill his or her complete responsibilities and to help everyone learn new ways of working together." This commitment to training and education includes providing the management- and executive-development programs at the GTE facility. The GTE experience was discussed in the previous chapter as an alignment mechanism, and it served that purpose in Didier's case. Listening to the human resources people and the salespeople in his management-development seminar talk about a business and market focus influenced him to become aligned with the new, comprehensive conception of Champion strategy and to understand his role in it. Nearly sixty separate programs are offered at GTE in general management, specific skill and knowledge acquisition, sales and marketing, instructor training, and organizational effectiveness. The general-management curriculum starts with supervisory skills for the new supervisor and moves through programs for middle managers to those for executives. Some of these programs are required for anyone who moves into a new level of responsibility, while others are "elective." Most of them provide a "back-home" action plan and a follow-through component. In addition to the programs at GTE, training and education in technical, social, and business knowledge and skills go on at all the mills.

By 1990 the Hamilton mill had become a company benchmark for well-organized, competency-based operator training, with operators themselves pulled off their regular jobs for a period of time to develop skill blocks and training manuals. This training model makes good sense on the face of it: involve the people who are experts in doing the work, thus building the competence these operators have derived from years of experience into the training of those with less experience. However, before this approach to training could become common practice across the mills, a fascinating bit of cultural change had to take place. In the late 1980s, prior to the tense adversarial bargaining with unions in 1989–1990, mistrust of management by the rank and file caused them to be concerned that the manuals would be used to train supervisors and other salaried employees to run the mills in case of a strike. Employees and union officials referred to them as "scab manuals." To protect themselves they refused to participate in the development of such manuals even though the lack of solid skills training was commonly acknowledged and decried. It took overcoming the bitterness of the 1989–1990 contract negotiations without a strike or lockout and the subsequent period of trust building on many fronts before this fear began to recede. Now operator involvement in competency-based training is widespread. (Labor-management relations will be dealt with more fully later in this chapter.)

In addition to technical training, operations and maintenance employees received training in interpersonal and team skills such as problem solving and communications. Such training is most effective when it is done at the time a work group has need of it rather than on a mass, across-the-board basis. Often, when employees go through training in skills months in advance of using them, the skills are forgotten; just-in-time training provides the opportunity for immediate, real-world practice and reinforcement.

There is growing interest in the management literature in open-books management. The previous chapter pointed out Champion's open sharing of information in its efforts to create alignment. Employees need more than just information however; they must know how to understand and use it. Several mills developed programs in which they explain the economics of the paper business, what the cost levers are, how budgets are developed and managed, and similar topics. The emphasis is on employee line of sight (employees are able to see the impact of their performance on organizational gains) and line of influence (employees are able to influence organizational gains through their performance) in contributing to business effectiveness.

Supervisors were trained at the mills in leading teams and boundary management—that is, managing the environment that surrounds the work teams rather than the throughput process. Supervisors, shop stewards, and others were trained in interest-based negotiations or problem-solving methods and in applying the principles of mutual-gains bargaining to resolving everyday conflicts. Teams at all levels were trained in collaborative skills to help them with interpersonal communications, conducting effective meetings, and situational leadership. A program in modes of decision making helped groups determine which of a variety of ways of making decisions, from totally hierarchical to totally democratic, is the most fitting for a given issue. The concepts and skills of Total Quality Management were offered. Many companies have provided similar programs as they have scrambled to become or stay competitive in today's turbulent business environment. Also as in many other companies, Champion locations were sometimes guilty of treating these programs like flavors of the month or stand-alone programs into which employees were "sheep-dipped." Fortunately, because of the larger systemic change process, these programs more often became an integrated part of a location's strategy, customized to fit the particulars of that location's culture.

Learning was not confined to the classroom. A great deal happened on the job through effective performance feedback, team visits to customers, and intentional career-management initiatives such as transfers and promotions.

Although training and education were emphasized throughout the company, the mill in Sartell, Minnesota, demonstrated a particularly strong "education ethic." Perhaps as part of a state or regional culture, employees of that mill place an unusually high value on their education. Sensitive to that aspect of the mill culture, whenever the joint union-management leadership group wanted to introduce a new part of their change process, they were quick to offer a training program on the subject to all employees. They also carefully track participation in all their training programs to ensure they are targeting the right people. Opportunities for personal growth are among the most rewarding factors in a job for many people, although, unfortunately, this factor is often overlooked. The Sartell folks, however, clearly recognize its importance. As a result, not only are people skilled and knowledgeable, but, of perhaps equal importance, their self-esteem is enhanced. Thus they go about their work with confidence as well as competence—a high-performance combination.

Growing confidence through growing competence carries over outside the workplace as well. An important but seldom appreciated by-product of the development of people in high-performance systems has been their increased sense of responsibility and leadership outside the workplace as they realize these potentialities in themselves at work. This transfer of confidence was true at Champion locations as well.

Reinforcing Learning, Cooperation, and Commitment: Rewards

We discussed the role of rewards and recognition in creating alignment in the previous chapter. Here we examine how they affect employee commitment. Barry Macy is the director of the Texas

Productivity Center, which maintains a data bank called the Survey of Innovative Organizations based on one hundred nontraditional organizations in North America. The average plant in their survey, Macy says, has two pay systems, such as pay for skills plus gainsharing.[1] Some have three, adding, for example, profit sharing. We have previously mentioned some of Champion's initiatives along these lines: pay for skills (growing out of redesigns to encourage flexibility through multiskilling), gainsharing, and a companywide program instituted in 1995 called Something Extra. We examine some of the effects of these initiatives in this section.

Pay for Skills

The jury is still out on pay for skills at Champion. Under a pay-for-skills system, employees can qualify for higher rates of pay by acquiring additional prescribed skills. They need not exercise all the acquired skills on a continuous basis but are able to handle the additional responsibilities when needed. Usually, there is not a cap on the number of employees in a team who may qualify for higher rates of pay. This system contrasts sharply with evaluation and classification schemes, and the extra cost is justified on the basis of expected increases in flexibility, motivation, and cooperation. Macy reports pay for skills associated with improvements of 40 to 70 percent in quality, cost, service, and delivery. Because of support work picked up by operators through multiskilling, companies in his survey report natural attrition of about 33 percent. However, there have been cases where pay for skills is poorly administered, skill certification is done sloppily if at all, employees are almost automatically approved for the next skill and pay level, and companies end up paying for something they don't get. One academic who consults frequently with trade unions doesn't believe that pay for skills works at all but advises unions to accept it if offered because it is "virtually more pay for nothing." Champion's experience probably encompasses this whole range.

For example, one of the paper mills did a thorough review of all of the areas that had been redesigned over about a nine-year period. When they looked at job rotation, a common practice in pay-for-skills programs, some employees reported that it was one of the key aspects of their design that was working well. They said it increases awareness of the effects one job has on others, decreases routine work, and helps in the leveling out of work loads when people are absent, including during peak vacation periods. "Most people do their jobs," they said. "Rotation of jobs is good. No one's locked into just one function of their jobs and that decreases burnout and boredom." However, other employees in the same area cite job rotation under the heading "What's Not Working." They said that rotation leads to "no one taking ownership in their jobs." On the one hand, some said pay for skills increases flexibility and "gives bottom jobs more money sooner than traditional pay practices." On the other hand, others said that "duties are constantly being added with no pay increase," and people lose motivation when they "top out." Another department praised the advantages of flexibility in scheduling and its ability to be "adaptive to rapid change in markets" as a result of multiskilling, and another said that "continuity between shifts and jobs [has] improved tremendously."

Some of the pay-for-skills schemes divide groups into "job families" with the restriction that workers can move through a "gate" to a higher family only when there is an opening rather than simply when they have acquired the higher level of skills. This scheme also has drawbacks. The report from one area said, "There exists a great deal of friction between the three job families, some resulting from pay, some from status, some from job duties. [This is] no different than friction between jobs in a line of progression; however, the three groups are large in numbers for each family. Therefore any one group can stir up a big stink all on their own as one's perceived misery can find lots of company." In other words, this scheme takes the typical enmity derived from traditional hierarchical pay schemes and elevates it to a new level as a result of intergroup dynamics. Polarization among groups who must work together is likely to be greater and to have a graver impact than the polarization that occurs among individuals in traditional settings. Yet the report says that the output in the same department where these conditions exist is the benchmark of the paper industry in terms of production, cost, and quality.

What conclusions can be drawn from this mixed bag? Perhaps pay-for-skills efforts have significant potential for producing both positive and negative impacts. Pay for skills is a powerful tool either way and must be designed and administered carefully, keeping in mind issues such as those above. In these reviews are evidence of Frederick Herzberg's famous "hygiene/motivation theory," in which pay for skills leads to learning, growth, achievement, and recognition—all strong motivating factors—and dissatisfaction with pay and with the administration of the policy are the hygiene factors.[2] The flexibility of skills and the advantages that flexibility brings are difficult to imagine without a compensation framework along the lines of pay for skills, and Champion is continuing to review and fine-tune its practices in this area.

Something Extra

In 1995, as the industry emerged from a deep trough, Champion and the UPIU announced a one-time, all-employee incentive plan called Something Extra. The plan was conceived by senior managers who saw the potential for 1995 to be a spectacular year from a revenue standpoint. Enhanced production levels would produce financial rewards, and this would be a great time to share with employees the results of their efforts. Employees at every domestic operation would participate, based on improved production at their own location and the performance of the company as a whole. The goals were aggressive, as they were being put in place at a time when record performance was already being achieved at many operations. But more potential was there, along with the desire to attain it.

There was an isolated precedent for this action. A major pulp-production improvement project had started up in the spring of 1992 at the Courtland mill. During the interim between that start-up and the start-up of the #35 paper machine in Courtland in the late summer of 1993 (Chapter Two), the mill was faced with an oversupply of pulp. Among the ideas for how to use this temporary abundance was the desirable option of producing additional paper with the existing machines in the mill, thus drawing more pulp from the new pulp system while increasing the incremental tons of paper production, all of which would have a beneficial effect on the return on existing mill assets. Mill management, with sanction from Stamford and in consultation with the local unions, decided to introduce a simple millwide incentive based solely on quality tons produced along with some safety requirements. It was made clear that this was a temporary add-on incentive plan that would terminate with the start-up of the #35 machine. Moreover, the expense was not great. The plan worked well; production records were set nearly every quarter, and bonuses totaled over $1,000 per employee. No repercussions were experienced when the program ended; in fact positive trends continued after the start-up of the new machine.

Champion leadership had studied and was at that time encouraging mills to investigate and adopt fully developed gainsharing plans, but it was becoming clear that such plans would be adopted only mill by mill over the next several years, at best (see next section). Although the Courtland model was not a pure gainsharing plan, lacking the subtleties of a balanced formula as well as the ownership that participation by employees in the plan's design could bring, it did begin to lay the foundation for employees' acceptance of and readiness for gainsharing. Sigler believed that a companywide incentive plan such as Something Extra could be an important next step on the road to gainsharing across the company. Corporate executives and some mill management were concerned about instituting a program that was based only on tons produced. It could be a setback, they argued, in the efforts to change a perceived "tons are us" mentality to a comprehensive, systemic, and customer-focused view of the business. However, some believed that this broad, complex view of performance was sufficiently in place and that company leaders could continue to make clear that the overall priorities had not changed. Besides, the argument went, it would be impossible to develop exactly the right formula to cover the special circumstances of all locations in a companywide program. Moreover, even if it were possible, there wasn't time to do it to get the jump start desired for 1995, and it would preempt locations from eventually developing their own gainsharing plans. Tons were a common denominator, and business conditions were such that additional incremental quality tons would indeed have a major positive impact on the bottom line.

In the end, it was decided to take the leap of implementing a Courtland-like incentive plan across the company, approaching it in an experimental frame of mind, aware of the risks and ready to modify or pull back as experience dictated. Again it was made clear that Something Extra was indeed just that: meant to last for only one year and not intended as a permanent part of the compensation package. The plan was sanctioned by the UPIU on that basis and jointly announced.

Something Extra goals were set for each mill based on a degree of stretch over and above the production targets they had already set. Goals for corporate and business-unit staffs and all salespeople were tied to the performance of the eleven paper mills. Forest-resources locations' goals were based on the paper mill they supported. The companywide goal was based on the total tons per day of the eleven paper mills. On a quarterly basis, if a location met its goal, each employee would receive $200 for that three-month period. If the companywide goal was met, each employee in a mill would receive an additional $50, for a potential quarterly payout of $250. Individuals in staff groups would receive their $250 if the company goal was met.

The plan was initially received with mixed feelings in the mills. No one wanted to turn down more money, but many of the same concerns of the executives were raised again at the local levels. Some in both the unions and management felt that corporate leaders, whom they credited with being so steadfast through the down cycle with their persistent support of the change effort—including training, safety, quality, and customer focus—were now forsaking those priorities under the pressure to make money in good times. Unions were especially concerned about losing important gains in safety. They feared the return of what they saw as the worst side of local management—a narrow focus on production at all cost.

Mill managers shared some of these fears, which were compounded by a perceived threat to their personal employment security. Throughout the downturn Stamford leaders had supported mill management, applauding and encouraging their successful efforts at continuous improvement even though depressed prices resulted in losing money at almost all locations. Now, some local managers felt, the jig was up; no one could hide behind low prices any longer. It was perform or else, and Something Extra standards for each mill, consistent with the program name, were set even a bit higher than the official operating plan. In addition, mill performance against those standards would be much more visible—frequently published, distributed on a list with the results from other mills, and posted on bulletin boards all over the company. Never before was mill performance so widely publicized. What if you didn't make your numbers? Everyone would know. It would be embarrassing to say the least, and, more important, it could negatively affect your career or even cost you your job. Some people in the mills felt like members of a professional football team that had successfully fought its way to the Super Bowl by being aggressive and taking risks, but, now that they were in the big game, they were tightening up and forgetting what got them there. Instead of playing to win, a part of them wanted to hunker down and play it safe in an effort not to lose. Certainly not everyone experienced these fears. Some were eager for the challenge.

By the end of February (the second month of the program) results were posted all over the company. Even though standards were intended as a stretch over and above plan, nineteen of twenty-five locations met or exceeded their goals, and all the others were close. Executives also took pains in the first months to explain and clarify the intent of the program and, as balance, to stress other top priorities for the year, including their continued support of the overall values and goals of the change process. Fears in most places eased, although pressure was felt by those mills that had not been so successful.

As payout time approached near the end of the first quarter, energy was high in mills that were close to meeting their goals. They worked hard to see that they closed any gap or prevented any last-minute slips-ups. However, the program didn't receive as much attention in the mills that were well over the mark or that had no chance of achieving their goals. This result is consistent with other findings on this aspect of the motivational value of goals—that is, goals are generally likely to motivate only when there is something like an even chance of meeting them; they are not likely to motivate either those who see only a slim chance or those who will easily surpass the goals.

All but three locations met or surpassed their first-quarter goals resulting in 102 percent achievement of the companywide goal. Thus everyone received at least $50 while most took home the maximum $250. Although the bonuses did not amount to a great deal of money, the checks were an important source of recognition, and people took satisfaction in their mill's achievements.

Because the program went so well the first quarter, it was decided to add to the challenge by "sweetening the pot" a bit for the second. Calling the new feature Improvement Plus, the company added $25 for each percentage point over first-quarter performance on top of the regular payout for achieving the goals set for the second quarter. All but two locations made their goals, with companywide performance in excess of 103 percent of target.

As the program progressed, there was a settling-in process, and it was no longer viewed as threatening. Most people seemed at least to accept it as a generally positive program. It still added somewhat to the pressure to produce, but it was also rewarding in the vast majority of cases, providing tangible recognition in each mill and in the company as a whole for a job well done. The fear factor was mostly gone.

"When we first started Something Extra in January," said the senior vice president of organizational development and human resources, "we never dreamed we would be at the level of sustained performance we've been able to attain." After inclement weather and other obstacles caused a difficult July for a number of locations, most rallied and turned in positive third-quarter results. In November the Congratulations Cash program was added for mills that exceeded their fourth-quarter goals by 1 percent and that had also achieved their Something Extra goals in any other quarter. Each mill employee would receive $50 if his or her location had met its goal once, $100 for two quarters, or $150 for three quarters. The total possible payout for the fourth quarter was $400.

As the program neared its end, the consensus was that Something Extra had met or even exceeded the most optimistic expectations. "It's amazing since many of us were sure that it would fail," said one senior executive. Almost 75 percent of a companywide sample of employees surveyed said they either agreed or strongly agreed that Something Extra had positively affected company performance. Forty-one percent said that it had positively affected their own performance. Even before the final results were in, discussions began about how to follow up; what, if any, program should take its place for 1996? There was now the risk (acknowledged from the beginning) of the "What have you done for me lately?" phenomenon. When the final results came in for the last quarter, with almost all locations exceeding their goals and receiving their Congratulations Cash, the announcement was made to continue with a new program for the coming year.

From the beginning almost all the reactions from employees to Something Extra were the predictable ones. When a new program is announced, it brings with it much uncertainty. It is common for employees to entertain thoughts that upper management has lost its mind, even to question its competence. Worst-case scenarios are shared around the water cooler; self-protection efforts manifest themselves, including various forms of bargaining to escape any perceived threats. As the new program starts up, there is typically a sense of confusion and disorientation along with a great expenditure of energy, much of it wasted in wheel spinning. If the program is to be successful, people then move through the problems and find solutions, creating a sense of direction and reduced uncertainty. Ultimately they believe that things are better after all. In this respect, the one-year experience with Something Extra might be seen as a microcosm of the whole decade of change at Champion, the short time frame bringing these stages in individual response to organizational change into clear relief.

A legitimate issue for exploration might be what mattered more as a performance inducement—the money or the widespread publicity of results, with its potential for embarrassment on the one hand or recognition and pride in accomplishment on the other. If the publicity was more important, would it be possible to get the same or similar results with visible feedback alone? These kinds of questions were discussed on occasion, but no systematic attempt was made to assess the issue or draw any conclusions.

There was an element of swashbuckling in this whole episode, which may say much about the confidence that leaders had developed in the change process and in their organization. Knowing they had less than the textbook solution, leaders took a calculated risk, also knowing that they could change the plan on the run. As it turned out, the changes were only on the up side as new incentives were added along the way, the opposite of the rate cutting so notoriously connected with many piece-rate incentive plans. There was even a spirit of playfulness; it was okay to have fun with the program. It also, as had been hoped, helped build a readiness for mills to go ahead with fully developed gainsharing programs.

Gainsharing

A Champion task force investigated gainsharing in the late 1980s, well before Something Extra was tried. As one important consideration, they examined whether gainsharing was most effective as a "leading" or a "trailing" intervention. In other words, should an organization implement gainsharing to get change started or use it later on as an equitable reward after significant change and improvement have taken place? Although they found examples of success and failure in both kinds of applications, they concluded that the weight of evidence seemed stronger on the side of gainsharing as a trailing mechanism. In addition to supporting the change process, Champion also established early on that gainsharing would be used as a compensation tool to improve business performance, promote teamwork, reward positive behavior, ensure continuous improvement, and reinforce pay for performance. Gainsharing guidelines were distributed, and location management and unions were encouraged to look into it jointly and determine their readiness to develop a local gainsharing plan.

Since then, the readiness assessment has become more formalized. (See Appendix C.) The process includes questionnaires, focus groups, and consultation with outside stakeholders. A location assesses how well it is doing with the ten "elements of success" as shown in Table C.1 and identifies gaps with a corresponding action plan.

Based on a significant payout at Quinnesec, the first mill to implement gainsharing, and a little "what-if" calculating by employees of other mills regarding what their bonuses might have been had gainsharing been in place in their mill in 1995, union officials in several mills reported eager inquiries from their constituents. In the spring of 1995 officials from all the locals and some managers and executives engaged in a discussion of gainsharing as part of a UPIU–Champion Forum meeting (this group will be described in detail later in this chapter), and some interesting perspectives emerged. One local union president expressed mixed feelings. His constituents were naturally interested in the possibility of earning more money and he wanted that for them; but he was concerned that gainsharing would prove such a strong incentive for employees to see things the company's way that it would weaken their union views and ties. He saw it as too company oriented and not supportive of the fundamental values of the trade-union movement. These are similar to the concerns of another local president who opposed it for his mill. He had expressed his fears to his constituents that gainsharing would add to the vulnerability of the rank and file by causing them to think more like management; thinking in this way could lead to such abuses as cost cutting at the long-term expense of the mill and their jobs or to layoffs, causing all employees to have to work harder than was safe or healthy or both. Still another local president who had some experience with a successful variable-pay plan similar to gainsharing disagreed with these views. He saw that people in his union were working smarter and maybe not even as hard as they had worked before while still achieving record performance that resulted in significant bonuses. He saw more committed and turned-on members. Further, it was his view that gainsharing was not an incentive for workers to be more like management but a matter of equity, providing workers the opportunity to share economically in the improvements they were a part of making. As an equity issue he saw it as aligned with long-held union values.

Today, Champion has thirteen gainsharing plans. They have been implemented in four paper mills, one DairyPak plant, two wood-manufacturing facilities, five forest-resources regions, and sales and marketing for forest products. Three more plans are currently being developed.

Over the decade many key lessons were learned about this type of compensation. The following are just some of these findings as they relate to Champion's philosophy:

- Gainsharing must be aligned with the company's culture. It must be consistent with Champion's value structure and stated goal of seeking "the active participation of all employees in increasing productivity, reducing costs, improving quality, and strengthening customer service."
- A readiness assessment is an essential process that helps identify key gaps and determines whether gainsharing would support the change process.
- The plan must be designed with factors that are within the control of the plan participants.

- Plan participants must understand (a) the connections among individual effort, group effort, and the performance of their facility, and (b) how they will be rewarded for their actions (both individually and as a group).
- Gainsharing requires commitment and action from management and all participants.
- Gainsharing is not a substitute for competitive compensation. It is additional compensation for significant performance and improvements.
- A gainsharing plan should be self-funding.
- A profit threshold must be considered in all plans.
- Joint communication must exist between management and plan participants to implement and maintain the plan.
- Each plan must be custom tailored to the needs and goals of the group.
- Plans should have a sunset provision and be reviewed every year or two for renewed commitment and approval.
- Gainsharing is not part of the negotiation process with the union(s).
- Issues resulting from the plan are not part of the grievance process and are not subject to arbitration.
- Education about gainsharing principles, business basics, and the plan itself is critical.

Champion believes these plans have been successful. Although success tends to be measured in whether there are payouts or not, other measures of success are improved union-management relations, employee line of sight, employee understanding of the overall business, and employee involvement with customers. These are just a few of the positive changes often noted at locations with gainsharing plans.

Executive and Board Compensation

To reinforce a heightened commitment by CEO Dick Olson and his new executive team to increase shareholder value, top management and the board compensation committee increased the emphasis on return on capital employed and total shareholder return. Key managers now would be required to attain designated levels of common-stock ownership, and a major fraction of the compensation of the board of directors was changed from cash to common stock.

Other Rewards

Recognition is a reward factor in all these pay practices. Added money in the pocket is tangible recognition whether it marks individual gains in skill acquisition or group and organizational gains in performance as measured by a variable-pay formula. More symbolic forms of recognition were used frequently by Champion in the area of safety; t-shirts, pens, barbecues, and other tokens were given at the location level, and an annual corporate safety award was presented to those locations that had no lost-time accidents the previous year. However, this form of recognition was used little in other areas.

Perhaps most important in furthering the change process was the recognition that came to groups that successfully innovated. For example, the Courtland wood-yard design team developed and implemented a successful redesign of their area. The wood yard is where papermaking begins; logs and woodchips are received, stored, and processed on their way to becoming the wood-pulp fiber from which the paper is made. Beginning at the beginning like this was a sound tactical choice for the Courtland mill change process because this department's performance affected everything that happened downstream, and improvements were widely recognized by others throughout the mill. In addition, as the new design was being implemented, visitors from other mills and other companies came to see it in operation. This recognition helped the people in the wood yard feel even better about their redesign and encouraged them to continue to improve, which in turn brought more recognition. This example was repeated many times at many Champion locations; some mills had to put controls on the number and frequency of visits by outside groups lest they spend so much time talking about what they were doing they wouldn't have time to do it.

Nevertheless, this cycle of performance-recognition-greater performance–more recognition gave a boost to the continuous-improvement efforts. One result was a shift in the motives for change. At first a new initiative usually started because someone outside the group or higher up—such as the corporation, a mill manager, a union president, or an internal or external consultant—suggested it. As a small group of employees in the unit became involved, they began to take ownership. With successful implementation came the recognition of visible and solid performance, perhaps some economic reward, and visits by others who had heard of their success. Ownership became more widespread throughout the unit, and the motives for change were increasingly internalized. The employees now change because it works for them in a variety of ways, not because somebody told them to. Performance is improved, life is better, and commitment grows.

Revising the Employment Contract—Commitment Is a Two-Way Street

Commitment is a two-way street however. Employees must see that their commitment to high performance is matched by the company's commitment to them. Compensation programs such as those discussed help. However, in today's highly competitive global climate, massive layoffs have become common, even among groups and at levels that in the past were relatively immune from such threats, and the old notion of working for life at one company has diminished. Some observers have seen in these developments a dangerous abrogation of the social contract.

Champion considers itself dependent on the skills and goodwill of long-term employees. Many operators and maintenance employees have worked more than forty years for the company. Although there were fears when the change process began that it was going to lead to major workforce reductions, such was not the case over the decade leading up to the Gang of Eight's analysis in 1997. There was indeed some prudent trimming, but from the beginning Champion tried to allay fears through language (for example, the Champion Way in Action statement, which promises no layoffs as a result of work redesign) and behavior. Most telling, perhaps, during the difficulties in the industry in the early 1990s, Champion never once shut down operations because of slow orders even though competitors were doing so. Employees and their union representatives recognized and appreciated that practice. These factors would weigh in the Gang of Eight's decision-making process, as covered in the Prologue and Epilogue.

A Summary of Rewards Initiatives

Rewards can take many forms and play many roles in the transformation of a corporation. The changes themselves can enhance the attractiveness of the job, workplace, or organization—the quality of work life—and hence provide increased *intrinsic* reward for employees. Employment assurances or pay practices can serve as *inducement* to employees to accept changes that they might otherwise oppose or be ambivalent about. Demanding goals and thresholds for group bonuses can be used to create a *sense of challenge* and if met to generate *pride*—both feelings that can reward effort even when the tangible gains are more symbolic than material. Salary and bonus systems can generate the *motivation to excel*.

All these forms of reward and their effects were present in the Champion transformation. The multiplicity of these forms and their effects strengthened the change effort. In addition, the combinations and sequences in which they were brought into play also enhanced their effectiveness.

Reforming Union Relations: Toward Mutuality

We have referred more than once to the difficult contract negotiations of 1989–1990. Those years were ones of "touching bottom" in this transformation story, and it is important to understand them. To do so it is necessary to go back, even farther back than the beginning of the change process that we have dated to the mid-1980s, to a strike, a painful work stoppage at the Courtland mill in 1980. The roots of that strike were growing throughout the paper industry in the 1970s.

When compared with manufacturing generally, paper is a highly capital-intensive industry in which labor costs are a relatively small percentage of total costs. Therefore it is critical to keep the huge pulp mills and paper machines running near capacity without interruption. Although unions in the industry, with its rural origins, did not have a history of militancy, they did indeed take advantage of their leverage, and wages grew faster than in other large manufacturing industries in the 1970s. Historically companies negotiated at the local mill level, but a pattern had developed by this time of keeping pace with the largest company, International Paper, where annual raises were in the 10 percent range by the end of that decade.[3]

A major power shift—brought about in part by advancing computer technology, which made process control less dependent on experienced labor—began in 1979, when Crown Zellerbach and Weyerhaeuser shut down mills and reopened them with replacement workers to avoid agreeing to union proposals for pattern wage settlements. When the unions struck Champion in 1980 at Courtland, management chose to continue to run the mill with salaried employees manning the process facilities.

The impact of the 1980 Courtland strike would be felt throughout the company and its unions for the next ten years, ten years of mixed strategies and signals bordering on a kind of schizophrenia. The one moral all sides could draw from the strike was "There's got to be a better way!" No such consensus was found in answer to the question of what the "better way" should be.

Two streams to the story that unfolded after the 1980 strike were responses to this question. Sigler drafted the Champion Way statement as one answer. And it became the source of one major stream of the story: the attempt to foster more effective relationships and develop better alignment with the workforce. Much of the cultural-change effort described in this book can be traced to that source. In the first half of the 1980s this effort included an employee-involvement program and executive Champion Way visits to the locations. As we have seen, after the St. Regis merger this stream picked up momentum with new labels, new initiatives, and new energy from the top. Mill management was encouraged to involve local unions in their change efforts. Unions responded with caution everywhere; their response ranged from fairly positive about joint undertakings in some cases to outright rejection of the whole idea in others. Some locations made meaningful progress, but many struggled to the end of the decade. Sometimes the water in this stream had to be pushed up hill.

The second stream of the poststrike response to the question of what is a "better way" was to escalate the fight with bigger weapons. Discovering that it could run the mills (at least temporarily) without union employees strengthened management's hand (again, not only at Champion but throughout the industry). Led once again by International Paper, companies including Champion set out to make other contractual changes to improve their competitiveness, such as removing what they saw as restrictive work rules, holiday shutdowns, and premium pay under certain conditions. For example, Champion imposed so-called flexibility language in its 1985 contract with the Pensacola unions. This language lifted some contractual restrictions on the assignment of work to specific crafts and allowed operators to do a certain amount of their own basic maintenance. The word *imposed* is used here because the unions were caught without tenable strike power and were threatened by the prospect that the company, which had only recently purchased the mill from St. Regis, would not fund the needed facilities conversion that would enable it to move from making brown to white paper. These concessions at Pensacola set the pattern for negotiations over the next two or three years with the other mills.

The unions, however, were not about to yield without a fight. During the round of bargaining over work rules in the mid-1980s, the UPIU had begun to try to bring the locals together into a "pool" arrangement in which no local would settle until they all reached agreement. Even though each local negotiated separately and with different contract expiration dates, pooled bargaining would, it was argued, provide a means of collective action on a scale similar to companywide bargaining and would make it possible to strike a large pool of mills at once, making it much more difficult if not impossible for the company to run with salaried employees. This strategy was not mature enough during the mid-1980s to have much effect, but the UPIU was better prepared the next time around, when the most controversial substantive issue was the proposed elimination of Sunday premium pay.

The corporate employee-relations staff, charged with negotiating these contracts at most of the major mills in 1989, met offsite in January 1988 to develop a vision of their function in the year 1993, five years in the future. They foresaw a "shift from essentially an adversarial relationship to a participative relationship"; they considered the 1989–1992 transition as one of "high momentum with more of a universal focus on cooperative union/management relationships." "Contract negotiations," they said, "will be conducted in a participative mode where mutuality of interest will be the dominant force fostering new approaches to the resolution of mutual concerns. . . . More people will be involved in the negotiation process and there will be diminished discussion of economic reprisals, i.e., strikes, lockouts, employee replacement." All this language, of course, represents the first-stream strategy. In marked contrast, however, they predicted that the 1989 negotiations would be difficult, "conducted essentially in the adversarial model." In other words, although enthusiastically committed to a vision of cooperation, they, in essence, decided that it would have to wait until after the 1989 round of bargaining because they saw no way of avoiding a serious clash.

A short time later in 1988, a companywide management task force on union-management relations was formed to look at the future in this area as well. The powerful membership included, in addition to the employee-relations staff, mill managers from all the business units, along with department heads of mill operating and support departments. Reflecting its importance, it was actively chaired by Bill Burchfield, executive vice president of the printing and writing papers business unit, with consulting and facilitation by two of this book's authors (Walton and Ault). The split personality of the company at that time, however, resulted in a charter for looking at the future only *after* the 1989 negotiations, the assumption being that contentious negotiations were inevitable. This group, like the employee-relations staff, tried to focus on a first-stream future, with a "relationship . . . based on mutual trust, honesty, respect and dignity [in which] the parties recognize the mutual advantages of a high *commitment* organization." Try as they might to keep their deliberations focused on this goal, though, occasional disputes arose between operations managers, who were already experiencing the hostility that the Sunday-premium issue was generating in their mills, and the corporate members of the task force, who were charged with supporting and carrying out the power-bargaining strategy. Finally, as 1989 came and the difficulties of carrying on negotiations and operating the mills increased, the task force temporally suspended its meetings. As it turned out, it never met again.

In summary then, the parties dug in as they prepared for a tough year. No one knew then that one year would grow into two and that the final settlement would not be reached until November 1990. Walton and his colleagues tell the story this way in their book *Strategic Negotiations*:

> In three mills, management reached settlements with local union officials, and the agreements were ratified by local members. However, the UPIU succeeded in keeping the other four mills with expired contracts in the pool into 1990, creating impasses in these locations, whereupon management implemented the terms in its final proposal that were favorable to itself (but not those favorable to labor).
>
> In early June, 1990, the members of the locals at these four mills voted to authorize a strike by majorities ranging from 82 percent to 95 percent. However, the UPIU soon accepted the company's proposals with minor modifications. Both sides were undoubtedly relieved to have concluded a long and acrimonious struggle, but neither celebrated its achievements. Labor had succeeded in making its feelings felt and in demonstrating that it could coordinate bargaining across many mills, but it had not denied the company the major proposed changes that were in contention. Management had achieved its major substantive objectives, including the elimination of premium pay for Sundays as such, but the struggle had delayed for one to two years its efforts to move toward commitment and cooperation in most of its mills and had adversely affected performance in these mills for more than a year.[4]

This adverse effect on performance included severe drops in paper production in such key operations as Sartell and Courtland, the last two mills to settle. The union had been able at least partially not only to meet its objective of pooled bargaining but also to coordinate a slow-down strategy that in many countries would be called a "work-to-rule strike."

It had become obvious well before the final contract signing at the Courtland mill that the struggle was mostly about power, while Sunday premium pay was simply the line that had been drawn in the sand. At Champion (as well as in the paper industry generally), although the unions successfully exercised leverage through the 1970s by bargaining wage rates higher than those in other manufacturing sectors, the power balance shifted with the Courtland strike, in which the mill was kept running. Management then used its power to impose what to the unions were unsatisfactory settlements in the early to mid-1980s. In this sort of escalating fight, however, neither side is likely to surrender and give up what it sees as its institutional integrity. So the unions found a way short of using the severely weakened weapon of a strike—coordinated bargaining and working to rule. The company could not be seen to cave in either and held out despite suffering losses of production until it finally "won" on such issues as the Sunday premium. Neither side really won; it was a classic example of lose-lose.

Once again in 1990, as after the settlement in Courtland ten years earlier, there was a clamor to find a better way. This time, though, there was a difference, and the solution found was to reverse the power-bargaining cycle by looking for a way for both sides to win. The first stream that had initially emerged from the 1980 Courtland strike in the Champion Way statement had reached enough momentum to take over and become the unified strategy of the future. Danny Morris, president of Courtland's largest local, and Bill Bannan, the mill operations manager, jointly vowed "never again" and launched a change strategy—beginning with the redesign of the wood yard—that resulted in the success stories recorded in Chapter Two and this chapter. In a similar partnership Tom Dougherty, president of the Sartell local, and mill manager Bill Jordan worked together to lead a remarkable rise from the relationship ashes in that location. Frank Graham had been a successful innovator with people as an operating-department head at the Lufkin, Texas, mill and was moved to the position of organizational-development and human resources manager about a hundred miles down the road in Sheldon. Shortly after his arrival he publicly committed to make 1994 negotiations at that mill a "nonevent." He and Jim Moore, the local president, forged a partnership and created a joint "leadership-skills" program that eventually involved every employee and manager in the mill. One by-product of this effort was the successful win-win bargaining for the 1994 contract. These are important examples of how the visions developed in pre-1989 by such groups as the corporate employee-relations staff and the task force on union-management relations began to take shape at the local level after 1990. There are many more examples throughout this book.

The shift also began at higher levels. In April 1991, still painfully aware that there had to be a better way, Presidents Wayne Glenn of the UPIU and Whitey Heist of Champion signed a Joint Statement of Principles, which read

> We commit to a relationship governed by the following four principles:
>
> 1. The UPIU accepts the responsibility to work with Champion's management to improve the economic performance of the enterprise in ways that serve the interests of workers, customers, stockholders, suppliers and communities.
> 2. Champion accepts the legitimacy and the institutional integrity of the UPIU.
> 3. Champion and the UPIU both accept the principle of increased worker involvement in creating a more participative work place in order to continuously improve quality and both customer and worker satisfaction.
> 4. We encourage management and union at all Champion locations to work together to achieve these goals.
>
> We believe that adherence to and advancement of these principles will benefit the Corporation, the Union, and all UPIU members of Champion International.

This step in turn enabled a further step a month later when the UPIU–Champion Forum was created. On the union side the Forum was made up of all the local UPIU presidents, a member of President Glenn's staff, and an international vice president. The company was represented by the executive vice presidents of the three paper business units; the corporate senior vice presidents of organizational development and human resources (coauthor of this book, Childers), health, and environment; and the vice president of employee relations. At its inaugural meeting, the group developed a mission statement: "to come together . . . to find ways to better understand each other and work together, . . . to exchange ideas and learn from one another, . . . and to promote continuous improvement of the change process with the goal of creating a more satisfied work force through greater employee participation, thereby improving the company's performance and creating greater employment and economic security."

This group continues to meet semi-annually; all but two of its meetings have been held at the mills on a rotating basis. The first day of a two-day meeting generally highlights the host mill, providing management and the local with an opportunity to tell the story of their change process. In preparation for their presentation the local parties reexamine and refocus on what they are doing, the progress they have made, and the gaps that still exist between the promise and the reality. The second day of the meeting may be on any of a variety of topics of joint interest. One marker signaling maturation in the relationship came in the presentations made at a meeting in early 1996. These presentations, on such topics as the UPIU and Champion positions on pending federal legislation and retirement planning, were done on a joint basis by staff representatives from the international union and from the corporation. In truly collaborative fashion, the presenters demonstrated the common ground the parties shared on these issues while respecting the differences they might legitimately have on some matters. In almost all the meetings there is an "open forum" period where members can bring up any issues or topics for discussion they care to as long as they do not involve bargaining. Again, these discussions grew in their openness and degree of mutuality. It became increasingly difficult (as well as less relevant) to ascertain whether the person speaking was from the union or management. This is not to say that differences were avoided or smoothed over, only that the differences could not necessarily be related to the speaker's affiliation.

Formal contract bargaining was also transformed between 1990 and 1995. With one exception—Pensacola negotiations in 1988—the talks between 1985 and 1989 were all totally adversarial and traditional, although not hostile in all cases. In the midst of the turmoil of the 1989–1990 round, there were two small lights at the end of the tunnel. While the Courtland mill was going through its painful process, the nearby timberlands unit (now known as forest resources), which supplied wood to that mill, quietly went about reaching agreement through interest-based, or mutual-gains, bargaining. Although the unit was small compared with a mill, it carried leverage beyond its size in that the workers there were represented by Danny Morris's local, the largest of the locals at the mill. This relatively small success contributed to Danny's commitment in the mill to a "better way" after the bitterly disputed 1990 contract settlement. A second glimmer of hope came from the bargaining at the Bucksport mill in 1989, the first of the paper mills other than Pensacola to try interest-based bargaining. The parties would later admit that it was far from a pure process, and they frequently reverted to traditional techniques. Nevertheless, they settled relatively peacefully in comparison with what was going on elsewhere, and the bargaining was cited throughout the company as a success for the win-win approach. The Canton mill followed that same pattern soon after. Post–1990, as unions and management at other mills prepared for their local negotiations, they often made joint visits to Bucksport and Canton to learn from these experiences.

The term of the 1989–1990 agreements was five years as opposed to the usual three. An employee at the Courtland mill was quoted as declaring, "The company forced a five-year contract on us. We can live with it. Can they?" Although the employee clearly intended the statement as a threat, others could find other meanings in the same words: opportunity, hope, or challenge. In fact, the interest-based approach to resolving conflict grew and matured rapidly over those five years. The concepts and skills were taught widely throughout the company. The parties in many cases decided jointly that these competencies were applicable to more than formal contract negotiations or bargaining, and they began to call the approach "interest-based problem solving." In many mills first-level supervisors and shop stewards were taught these skills; they were asked to apply them to resolving issues before they became formal grievances as well in the formal grievance procedure itself. By the time the five year contracts expired, interest-based problem solving was the norm, and all new agreements were reached in this way. The two most rancorous negotiations in 1990 were at Sartell and Courtland. In 1995, Sartell reached a win-win agreement in twenty-two days and had it ratified and signed well ahead of the expiration deadline. Courtland's contract, reached this time on a mutually agreed basis, was for an unprecedented six and a half years and was ratified by 76 percent of the rank-and-file voters.

Of even greater importance, however, was the amount of progress that joint leadership made at the local level on changing the climate of the organization from one of control to one of commitment. Joint leadership teams were formed at each mill to sponsor and support change. The leadership team at Bucksport, for example, helped bring that mill back from the brink by focusing on total quality after their major customer had threatened to take its business elsewhere. The Deferiet, New York, mill, with some of the oldest, least cost-competitive facilities in the company, was a chronic dilemma for Champion. Should it be closed? Can it be sold? The joint leadership team worked to trim costs and downsize without one person involuntarily losing employment. Union leadership took an active part in targeting at many mills, sometimes making the trip with management to Stamford for the annual review of their five-year plans. As mentioned in Chapter Two, the joint leadership at Sheldon brought the targeting process to the crew level, involving every employee in the mill. In most mills, the joint leadership teams took on the difficult health-care issue and worked together with community groups (in some cases actually initiating community involvement) to develop locally appropriate means to contain health-care costs.

This kind of cooperation over the five-year period led to increased employee competence and commitment throughout the company at all levels. Perhaps, however, the most remarkable demonstration of this new level of overall capability was the win-win situation enjoyed by the Sartell mill joint leadership team in 1995, after a decade of change: (1) they negotiated a new contract in twenty-two days; within weeks (2) employees overwhelmingly voted for a jointly designed gainsharing plan and (3) cashed their Something Extra checks for a year of outstanding performance. As a pointer to the future, their new contract language formally recognized interest-based problem solving as the preferred means of resolving differences and negotiating agreements.

As another sign of things to come, in July 1995, UPIU Local 193 and the Champion Alabama Region Forest Resources Group signed what they called a "partnership agreement" instead of a traditional contract. These were the same groups that had been the first in the company to use win-win bargaining in settling their previous contract. This partnership agreement was only nine pages long and began with statements of "Our Joint Mission" and "Our Partnership Philosophy." In a section on "Operating Structure and Employee Participation" the agreement states that "work will be accomplished in the Region by geographically based, self-managed Work Unit Teams. These teams provide the basic structure in which the Region will implement our Partnership philosophy and our principles regarding employee involvement. Union and Management agree to seek full participation by all employees, utilize a clearly defined decision making process, provide for the free flow of information, emphasize competency and training, and place decision making at the most appropriate level." This paragraph virtually institutionalizes the entire "what works" formula in contract language. As mentioned in Chapter One, in recognition of this remarkable progress across the company, the American Society for Training and Development awarded Champion and the UPIU its 1995 award for "Outstanding Achievement in Employee Involvement in the Workplace."

Soon after, in a potent coincidence, the top leadership of both Champion and the UPIU changed in 1996. Sigler and Heist from the company and Glenn from the union all retired after long service with their organizations. Dick Olson replaced Sigler as CEO, and Boyd Young was elected to the presidency of the UPIU. Like Olson he was not unfamiliar with the notion of high-performance organizations with highly involved and committed employees, and he was supportive of effective partnerships with companies provided they were sincere in their intentions. At the first meeting of the UPIU–Champion Forum following their respective promotions, they emphasized their continued support, Olson in person and Young through his vice president and chairman of the union's Champion council, Don Langham. (Young, in a further sign of changing times, had been called to Washington, D.C., at the last minute by another new leader, John Sweeney, who had recently been elected president of the AFL-CIO.)

The Forum membership took the occasion to engage in a lengthy and energetic discussion of the future of the Forum and of the partnership as a whole. They expressed a strong interest in going beyond what they presently had and exploring opportunities for new breakthroughs. Olson participated with and encouraged them; it was a great opportunity, with new leadership of both institutions, he said, to reexamine everything and to chart new territory. As a result of these discussions, the members selected a joint group to plan a conference, with participants to include the Forum members and a larger group of mill managers and international union officials. The purpose of the conference was to take the first step in planning the next generation of the Champion-UPIU partnership for mutual success. This conference took place in December 1997, shortly after senior management rolled out its new business strategy (see Epilogue).

Summary and Conclusions

Competence is more than the ability to do the same job better; it is the ability to understand the system in which the job is located and to promote the performance of the larger system. Dennis Didier illustrated this powerful conception of growing competence—an idea discovered at Champion during the process of change, not one that drove the change from the outset. It should be part of the aspirations driving every corporate transformation. The Champion case is also illustrative of the many individual elements that can be tailored and combined to promote this system of competence— formal training, planning and budgeting, organizational redesign, reward schemes, customer-oriented campaigns, bosses as models.

High *commitment* and motivation are essential if greater competence is to translate into higher performance. The lesson underscored by the Champion story is that a high level of commitment can be generated by soft policies—such as participation and employment assurances—and then another level of performance motivation can be achieved by reward schemes based on results.

The evolution of the use of rewards in Champion's change effort is instructive. Champion relied heavily on intrinsic incentives to get initial buy-in and ongoing support for the change process during the first half decade. The rewards were generated by the participative process itself; individual employees derived increased satisfaction and self-esteem. Moreover, the Champion Way in Action statement sketched what many employees could accept as a "better way." But there was an important exception to this general pattern—namely, the pay increases associated with job redesign, a feature that clearly served as a tangible inducement for this particular type of change.

Then, with many of the changes already institutionalized and widely supported, Champion employed extrinsic incentives to more fully realize—and further stretch—the performance capabilities of individuals and the organization. The extrinsic incentives included the one-time campaigns—such as Something Extra—gainsharing, and tying the pay of managers and directors to shareholder value.

Having both intrinsic motivation and extrinsic motivation pulling performance in the same direction is a powerful combination. This fortunate condition can be created when intrinsic motivation is well established before extrinsic incentives are employed. In our experience at other companies, the reverse sequence can be highly problematical. If the initial change is generated or accompanied by extrinsic incentives, the tangible rewards provide the enduring primary rationale for accepting the change. It is then difficult to develop intrinsic motivation. Moreover, the motivation (based on extrinsic incentives) is readily demoralized if for any of a number of reasons monetary payoffs decline or disappear.

Cooperation, the third element contributing to organizational capability, refers to the idea that skilled and motivated individuals and units work together and not at cross-purposes. At Champion, both the social contract between workers and managers and the one between union and management had to be drastically revised. The challenge was one commonly faced by organizations with unionized workforces. The prevailing social contract with workers based on mutual compliance needed to be transformed to one based on mutual commitment. This transformation required that management change its own supervisory behavior—for example, toward involvement and participation—and employment policies—for example, promoting training and employment assurances. However, many of these changes were threatening to Champion unions, which had an adversarial relationship with management, because the changes were seen as having the potential to undermine worker loyalty to the union. Moreover, the unions could block or frustrate participation and other management policies to promote commitment. Therefore, management also had to change its social contract with its unions—replacing arms' length adversarial relations with mutuality—before much progress could be made in moving toward mutual commitment with employees. A major complicating factor, however, was that management also needed to change the substantive contract with labor; these changes increased competitiveness in the marketplace but disadvantaged employees (for example, by eliminating premium pay), and many of them were opposed by the union (for example, the elimination of work rules that restricted flexibility).

The derivative dilemma—again a classic one faced by management—is that promoting commitment and mutuality involves "fostering" attitude change, whereas achieving substantive changes opposed by employees and the union involves "forcing" approaches. Fostering and forcing approaches require mutually contradictory tactics. Forcing in particular can neutralize fostering initiatives. Nevertheless, research has shown that corporate managers who employ both forcing and fostering are more likely to achieve productive changes in labor relations and work practices than are those who rely exclusively on either forcing or fostering.[5] Champions experience—both in the pioneering changes at Pensacola in 1985 and in the diffusion effort in 1989–1991—were in line with this idea. Management employed both forcing and fostering—in that sequence—and in the end was effective in achieving both the substantive changes and the relationship changes it sought.

In brief, although building capability involves many initiatives, sometimes including apparently contradictory ones, skilled, dedicated people working together have the capability to do the impossible. But they need the appropriate latitude to exercise those skills, a subject we explore in the next chapter.

Chapter Four

Letting Go
Style, Structure, and Systems

*It's hard for managers to let go. I have been involved
where we were empowered but then found we were expected
to have done it the manager's way. Management has to
give us some room to make mistakes, and we have to give
management the same room for them to make mistakes.*
—Salaried employee, February 1995

*We had been accountable to strong, autocratic leadership.
Dick Olson, by his nature, was diametrically and
startlingly opposite. It was difficult to adjust to. Early on
I would get frustrated and say, "Damn it, Dick, make a
decision." But he suffered through. He has a great
willingness to listen and hear people out. One result is
that we all have a bigger stake in the outcome.*
—Member, Gang of Eight, August 1997

When our "what works" equation was first presented to Champion executives, a question was raised regarding which of the three factors was the most difficult to put into practice. Following an interesting discussion, the consensus was that although developing alignment and capability both offered plenty of challenge, appropriate letting go of authority was even more difficult.

Letting go of personal power and authority in order to gain increased systemic control is counterintuitive. It is a paradoxical notion that one can gain control of a situation by granting control to someone else. Compounding the difficulty is the fact that with letting go the goal is to optimize, whereas with alignment and capability the goal is to maximize. As Fred Emery put it, people need "adequate elbow room," so they do not sense that they "have some boss breathing down their necks, [but not so] much elbow room that they don't know what to do next."[1] Because of this ambiguity, managers often swing back and forth between providing subordinates with too much and too little discretion. In addition, programs or other initiatives to encourage letting go can be difficult to conceive because the goal is an appropriate *absence* of behavior (imposed control).

In this chapter we address these issues by looking at how Champion approached what we call the Three Ss of Letting Go: structure, systems, and style—each a mechanism for defining the appropriate latitude (see Figure 4.1). We report some of the problems related to leadership style with which Champion had to wrestle and how they were resolved. Letting go, however, is not just a matter of the style of individual leaders; an organization's culture as shaped by its structure and systems can range from imposed hierarchical control to control through employee commitment, and this culture, in turn, can reinforce or discourage particular leadership styles. We therefore include discussion of how the redesign of Champion's structures and systems contributed to empowerment at all levels. The combination of all three Ss is well illustrated in Champion's implementation of best-practices forums, the story of which concludes this chapter. We begin, however, with the subject of change management itself because the power as well as the dilemmas and struggles of letting go were perhaps nowhere else better illustrated.

Figure 4.1. The Performance Formula: Letting Go.

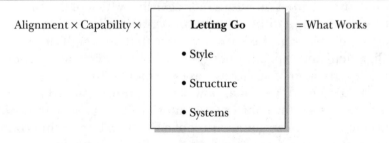

Alignment × Capability × | **Letting Go** | = What Works

- Style

- Structure

- Systems

Letting Go in Managing Change: A Matter of Style

When this process began, Champion was a centralized, function-ally structured organization. Human resources people in Stamford, for example, wouldn't allow mills to choose their own consultants for the change process. Training programs for the mills had to be developed or approved by training-department management in Stamford. CEO Andy Sigler spoke of the need for some group to be the "controllers" of the change process. By the same token, conventional wisdom holds that change must begin at the top and cascade down through an organization. Although such a top-down, across-the-board approach had its compelling logic for Champion executives operating out of their existing paradigm, it also seldom works in practice.

To make this point, a manager in the once-huge Fisher Body division of General Motors described what he saw as a common characteristic of his organization by putting his two hands together in the shape of an open cup and saying, "Our division doesn't resist new programs coming out of the corporate office the way some divisions do. We just take the program in [gradually closing the opening of his cupped hands] and 'Fisher Bodyize' the hell out of it. A year later, if you look in there [opening his hands just a bit to take a peak inside the cup], there's nothing left; the program has disappeared." This phenomenon is not limited to Fisher Body; it is common in organizations making efforts to change through new programs. Mass, across-the-board, "sheep-dip" strategies tend to spread energy too thin and are wasteful of limited resources as they are absorbed by the self-stabilizing system. Moreover, the best new ideas often don't originate at the top in the first place.

In contrast to the textbook, top-down approach, Beer and his colleagues researched actual "revitalization" attempts and found that those that worked "usually started at the periphery of the cor-poration in a few plants and divisions far from corporate head-quarters. And they were led by the general managers of those units, not by the CEO or corporate staff people."[2] Quinnesec and Pensacola served as the first such examples at Champion. However, these researchers also found that "having a CEO or other senior managers who are committed to change does make a difference, and when it comes to changing an entire organization, such support is essential. But top management's role in the change process is very different from that [in the classic top-down approach]. . . . Grass-roots change presents senior managers with a paradox: directing a 'non directive' change process."[3]

Such a paradox faced Champion leadership when the change process was launched in the mid-eighties. As covered in Chapter Two, Whitey Heist and the transition team were charged with central coordination of the effort, with Sigler a regular participant in transition-team meetings. Other "centralized" coordination mechanisms included the development and roll-out of the Champion Way in Action Statement, coordinated use of outside consultants, a corporate "managing-change" training program for change leaders, and a curriculum for internal facilitator-consultants. Two-day programs called "awareness seminars" were presented to senior managers in all the business units to provide an understanding of the nature of the changes being introduced.

Beyond these few measures, however, Champion came at its common goals from many different directions. No one program or method was introduced across the company; although not a good way to make changes, such a program would have made communications crisper and less ambiguous. The Champion change process was not strictly a sociotechnical redesign strategy or Total Quality Management or reengineering or any other of the fashionable approaches, although it contained elements of all of these and more. Even with the Quinnesec and Pensacola demonstration models, senior management recognized that a standard, cookie-cutter approach would not work in all mills, let alone other functions and departments across the value chain. Realizing that the culture and other situational factors were different at each location, senior managers charged the entire organization to move in the direction outlined in the Champion Way statements. Beyond those common elements mentioned above, each unit proceeded to develop its own strategies and to move at its own pace, with its own ups and downs in the learning curve. Some would call this general approach to change the plan-do-check-act cycle. Those less kind might call it trial and error or "ready-fire-aim." Many organizations would be uneasy with this approach and would want to adopt a standardized, top-down model. More than a few people at

Champion advocated such an approach from time to time. Sigler and others, however, successfully resisted all efforts at any sort of bureaucratized orthodoxy.

In the midst of a great deal of support and quality activity in the first few years there was also a great deal of confusion, anxiety, and wheel spinning. Champion's early experience was consistent with the typical characteristics of a transition period, predictably filled with high uncertainty, high energy, and high stress. There was continual griping down in the organization that management wasn't "walking its talk." Although this complaint persists to some degree, it diminished considerably in later years. A common response of management when confronted with this apparent chaotic state is to reach out to regain control, confirming subordinates' worst suspicions and resulting in an even wider word-deeds gap. As the uncertainty grows, everyone is likely to express some longing for "the good old days." These sentiments can be interpreted as resistance to change, but they simply indicate a desire for some stability. Again, the early stages of the change process at Champion followed this pattern. Nevertheless, in the words of executive vice president Dick Porterfield, the executives maintained their "dogged persistence and unjustified optimism." That meant continuing to provide what was sometimes an uncomfortable degree of latitude.

Actual change emerged largely in the field rather than at headquarters, in places like the Quinnesec and Pensacola mills, the Knightsbridge (Hamilton, Ohio) administrative center, and the West Nyack (New York) technology center, and was diffused to the company's other locations. These changes, even when successful, sometimes caused stress between corporate leaders and leaders of those locations, as boundaries were tested and issues had to be resolved. At one point in a 1992 transition-team meeting, Sigler raised the question, "Who's driving this thing? Because it sure as hell isn't at this level anymore. Control left here a long time ago." The correct answer at that time, if there was one, would probably have been that change was being driven largely at the level of the mill managers, although some drivers could be found both above and below that level. In later years, that would still be the case, with the addition of some corporate and business-unit drivers of cross-mill and cross-functional activities. By now the energy for change resided in many places throughout the corporation, and understanding of the change objectives was shared to such an extent that the top-level group could focus its attention on a few new initiatives, such as Something Extra and gainsharing.

Mill managers also often wondered about whether they had any control. More than one of them told before-and-after stories. Once upon a time, in the good old days, they might say, they could walk through the offices and the mill floor and know virtually everything that was going on, even feel in charge. Now they might wander around observing team meetings and other activities under way and not know who was doing what, even though they were confident they knew why—to improve the way the mill ran. It sometimes seemed like chaos. Bill Jordan, at the time manager of the high-performance Sartell mill, said he knew less about what was going on and felt personally less in control than he ever had and yet the papermaking process was more under control than ever before.

In follow-up interviews with supervisors and middle managers after redesigns had been implemented for several months to a few years, they were asked how their roles had changed. In general, they could only describe their new roles in terms of what they didn't do anymore or in terms of what their employees now did that supervisors formerly did. Most were unable to articulate what they themselves did that was different, what their new roles were. In such a void, it was perhaps inevitable that they occasionally reached back into the work group's arena and reverted to familiar controlling behavior. However, sometimes they let go too soon, before the other parts of the formula were in place sufficiently for employees to be "let loose." Managers at all levels, including executives, struggled with when it was appropriate for them to make a decision on their own, when to involve their employees, and when to let their employees decide on their own. They talked of "walking on eggshells," not knowing for sure how to behave. Employees asked for leadership on the one hand and, on the other, complained that managers weren't "walking their talk" if they didn't leave a decision up to the team. Training in the concepts and skills of situational leadership and modes of decision making were helpful, but perhaps nowhere else were the ups and downs of the change curve more evident than at every level of management (for that matter, union leadership struggled with some of the same difficulties).

With time and effort and lots of trial and error, however, modes of decision making became better understood and calibrated. Complaints continued that someone or some group had not been involved soon enough in a decision or that a manager made a decision without consulting his or her team. Gray areas will likely always exist. However, over time such complaints were much less frequent, and, by the same token, fewer managers expressed confusion about their roles in decision making.

Letting Go in Managing Operations: Forming New Structures

Redesign of structures played a major role in Champion's progress in letting go just as it did with respect to alignment and capability. We have covered this topic in some detail already, but here we want to stress that letting go is not merely one result of sociotechnical design but indeed is the central purpose of it. The whole concept derived originally from Eric Trist's observation of a natural experiment with self-managing work teams in a British coal mine; its first widespread application was in the Norwegian Industrial *Democracy* Project (italics added); and its basic philosophical thrust has always been "to achieve a shift in organizations from a bureaucratic structure to a democratic structure."[4] "The bureaucratic design principle," says Emery, "is that responsibility for control and coordination of work resides one level above where the work occurs. The democratic design principle is that responsibility for control and coordination of work is located with the people doing the work."[5]

We have described the realization of this shift at Champion at many different levels. To a large extent this was a bottoms-up development, starting with the redesign of work from specialized, individual tasks into multifunctional, shop-floor team tasks. Application of the same principle then moved up to product grade-line systems for mid-level mill management, mill leadership teams, business-unit "core business teams," and strategic teams at the top. Some of these groups were institutionalized in the formal structure; others were parallel structures, to which we add the many other forums in which people participated outside their daily roles. All reflect a shift toward enlarged decision domains for individuals and groups. They also reflect a shift from a vertical to a horizontal orientation.

In fact, as the new laterally oriented structures took hold at each level, the locus of power shifted to the level where the work was done, which was where the information, expertise, and motivation existed for getting that work done. Although some hierarchical control remains and perhaps always will, the shift meant that each level took on increasing discretion to manage its appropriate tasks or mission. For example, the role of mid-level mill managers became less a matter of closely supervising their subordinates (who, through redesign, were more self-managing) and more a matter of integrating the functions of an order-delivery system. Upper-level teams similarly managed themselves as they worked on strategic issues across the value chain.

With time teams also typically proved more capable of taking on these new challenges than individual specialists, thus making it easier for managers at each level to let go (more a case of pulling down authority than of driving it down). Strategic issues that once would have been handled solely, if at all, by a small number of senior executives with their functional hats on began, in about 1993, to be handed off to cross-boundary teams. Coordination of the sales, scheduling, manufacturing, and distribution of particular products, which would have been handled previously by business-unit executives, was delegated (cautiously at first) to middle-management core business teams. These groups, with the synergy of true teams, generally performed beyond initial expectations, and, as a result, more teams were formed to take on even more complex and strategically important issues. Business teams were institutionalized by the new executive team in 1996. Growing latitude, then, was a result of a cycle of pushing authority down, followed by a demonstration of increased capability, which resulted in pulling down even more authority.

Pushing authority down included some flattening of structures to fewer levels. Shortly after the Champion–St. Regis merger, a mill-effectiveness task force, of which Dick Olson was a member, recommended eliminating two layers of mill management: production managers (who functioned basically as assistant mill managers) and assistant superintendents. This reduction was implemented in 1986–1987. In the next few years, however, it was realized that these jobs had provided good opportunities for management development and that there was a danger of running low on talent. Therefore a program called Bench Strength was instituted in which high-potential mill-management candidates were actively recruited and placed in important coordinating roles but not added to the line hierarchy, leaving the reduced structure intact. Support-staff redesigns also resulted in the flattening of hierarchies in such areas as R&D (five levels of management to two), corporate accounting, and corporate facilities (five to three).

Letting Go in Redesign: Participative Processes

Although structural changes such as those described above came about in various ways, Champion adopted the sociotechnical model for most of their redesigns. The sociotechnical analysis and design process is itself participatory. Instead of management's deciding on a new structure and imposing it, substantial latitude is provided for all who are affected to have input throughout the process. A joint union-management leadership steering committee develops conceptual, policy, structural, and performance boundaries for the redesign and then sanctions a design team made up of a vertical slice of the organizational unit involved. Members of the organization have a part in choosing the design team, who then participatively go through a process of analysis and design. If they do their job well, they involve others in their unit often, soliciting information and ideas. The design that they come up with includes participation of all members in the ongoing strategic, coordinating, support, and process work of the unit. Employees are then involved in the approval and implementation of the new design.

As it turns out, even this level of participation is often not enough. Design-team members typically participate much more intensively in this process than others and subsequently develop much deeper levels of understanding and ownership. As a result, they often isolate themselves more than they should, leaving others out of the process too much. Thus, when it comes time for approval and implementation, their fellow employees are suspicious and resistant, sometimes completely rejecting the proposal. In essence, planning and doing have been separated once again. The fact that the planners are now fellow employees instead of management helps but a little and can create hostility toward the design-team members, isolating them even further.

The "accelerated-redesign" programs developed in the United States attempt to deal with this phenomenon through the use of large-group participative conferences. Champion locations have made some use of these methods. In fact, the accounting department at the Pensacola mill took this approach before it was even recognized as an approach. In 1988 they started out to redesign their twenty-person department. Because they were such a small unit, it didn't seem to make sense to pick out five to seven of them to function as a design team when that would be such a large percentage of the whole. Wasn't there some way to involve all twenty? It was decided that the total unit would meet in a series of workshops tied to the typical steps in the sociotechnical analysis process: (1) scanning the business environment, (2) analyzing the work processes or technical systems, (3) analyzing the social system, and (4) the redesign itself. A group of five was indeed chosen to lead this effort, but their job was not to do the analysis and design themselves but to plan the workshops for the larger group and to take the output from one workshop, fine-tune it, and use it as input for the next. The process worked well, and to this day that department is a benchmark for the application of the commitment model to office work. They came to call their process "the big bubble–little bubble" approach because when they drew it out graphically on a flow chart, the work of the large group was described inside a large circle and the work of the small group in a smaller circle. Approaches like this that aim to accelerate the pace of and increase support for the redesign and implementation process through greater employee participation are important and influential developments in sociotechnical redesign in the United States as well as in other countries.

Fred Emery, who, with Eric Trist, is considered a founder of the sociotechnical perspective, has advocated an even more participative approach to redesign ever since his return home to Australia after his experience with the expert-driven field-project redesigns developed in the Norwegian Industrial Democracy Project of the 1960s. Only in the 1990s, however, has this approach gained much favor and influence in the United States through the efforts of Fred and, especially, his wife, Merrelyn Emery, in seminars at New Mexico State University. A strong believer in participative democracy anyway, Fred was convinced that workers have the knowledge of the sociotechnical system in their heads and lack only the conceptual knowledge. He felt frustrated by the fact that, as he saw it, as sociotechnical redesign "achieved a high profile" in the 1970s and 1980s, "academics sought ways to scholasticize it [and] consultants sought ways to sell it," both adding unnecessary complexity to the process. "My focus has been to simplify the process of redesigning so anyone can understand it, use it, and re-use it."[6] The Emerys created Participative Design Workshops, in which in a day or two, depending on the complexity of the issues and the particular workshop format, an organizational unit can redesign itself. In contrast, the typical time span for doing traditional sociotechnical analysis and design is from several months to years. Participative Design is simple and straightforward enough, as well as short enough, for all members of a unit to participate without spending a lot of time off the job. There is no need for a design team to become isolated experts in the process.

Champion sent several internal organizational-development consultants and a few managers and union leaders to the New Mexico State seminars taught by Merrelyn Emery, and a few Participation Design Workshops have taken place at Champion with promising results. It was used with office workers in the Knightsbridge benefits office, with a transportation unit at the Canton mill, with the organizational-development and human resources staff of the Sheldon mill, and with the whole Champion organizational-development community as they structured themselves as a companywide "rapid-learning network" in late 1995. Perhaps the most ambitious application was alluded to in the Dennis Didier story. The entire management and salaried organization of the Pensacola mill redesigned itself in just four long days. Over ninety employees participated in a design conference; they were deployed for much of the time in four subgroups organized around the four business systems being designed.

Ironically the very advantages cited for these innovations in redesign methodology—accelerated pace and widespread participation—turn out also be sources of resistance to their diffusion. With the typical joint steering committees and design teams, both management and union leadership had learned to keep control of boundaries, most particularly boundaries around compensation. As long as only a small group—the design team—was involved and enthusiastically committed to what they created, then the steering committee, or even higher-level corporate management, could keep the lid on. However, management foresaw danger if large groups of employees became committed to designing structures outside the limits considered important by the sanctioning authorities. Further, when hearing about these new, speedier approaches, some of the same managers who had complained about how long traditional sociotechnical methods took began to say that a longer time frame allows everyone to deal with transition issues gradually. In other words, going slow gives management more control of the process. However these control issues come to be worked out, it is nevertheless true that work and organizational redesign at Champion is an increasingly participative process for coming up with designs for greater participation.

Other forms for participation around issues other than redesign have already been mentioned—for example, self-managing work teams, cross-functional middle-management teams and task forces, executive work groups, large and interactive targeting conferences. As with some of the other key developments in the overall change process, successes at lower levels of the organization eventually influenced higher levels, all the way to the top. Champion held an annual management meeting in October for years. One way it has evolved toward greater participation is simply in the numbers and levels of attendees. Until 1991 the group consisted of eighty to ninety senior managers, heavily weighted with corporate executives from Stamford and the Knightsbridge corporate staff groups in Hamilton, Ohio. The only attendee from a mill was the person at the top, the operations manager. Beginning in 1991, the group expanded to over 130 by adding several sales managers and three people from each mill beyond the operations manager. Nevertheless the meetings were fairly traditional, large-group executive conferences: seating was auditorium-style, and most of the agenda consisted of one senior executive after another reviewing the past year's performance, with an outside speaker lecturing on the ideas from his or her latest book. Once in a while there would be a cautious and tentative stab at using subgroups for discussion, but these were always seen as fairly marginal to the main business of the meeting.

Comparatively speaking, the 1995 meeting was revolutionary. Seated in subgroups for the entire meeting, participants engaged in an ongoing flow back and forth from front-of-the-room presentations to participation in groups to open-forum discussions in the large group. The focus was the future not the past, and the theme was "becoming the best" (see the Epilogue). Sigler functioned as facilitator cum talk-show host throughout the three-day meeting. Most participants declared it the best annual meeting ever, and they were encouraged to go even further in the direction of open participation in the future.

Letting Go in Managing People: Innovative Human Resources Systems

These moves toward a horizontal orientation and increased participation required redesigns not only of structures but also of the systems that support them. We discussed changes in reward systems in detail in Chapter Three, but other human resources systems were also revamped to increase latitude.

Modifying the Meaning of Advancement

With the shift to a horizontal orientation, employee concerns naturally arose about limited career-path options. People were accustomed to having opportunities to move up, but there were now fewer positions to move into. The whole notion of moving up and career ladders reflects a vertical framework, a mental model that had to be modified at Champion so that lateral moves could be seen as more significant than they were in the past. This change started with skill-based pay for shop-floor work teams and came to encompass staff groups in such areas as accounting, R&D, and facilities. Perhaps the highest level at which roles were rotated to promote multifunctionality was in the Brazil operations, where, for example, the senior accounting executive rotated into the role of senior engineering manager and vice versa. Titles and work roles such as team leader, process coordinator, facilitator, and boundary manager began to be used in place of supervisor or manager. These changes come slowly, and the vertical model was not and should not be completely displaced, but a shift is occurring.

Involving Peers in Performance Evaluation

Peer and "360-degree" performance-evaluation systems were also coming into use, starting with groups such as corporate facilities, accounting, and management information systems at the Knightsbridge corporate staff offices in Hamilton, Ohio, and moving to all management levels at some mills. One group at Knightsbridge, for example, designed their own process for 360-degree evaluations by creating a structure they call the "Performance Appraisal Team" made up of four members (management and nonmanagement) who rotate annually. This self-directed team gathers all feedback from anyone who has knowledge of an employee's performance, including peers, subordinates, managers, and customers; presents the feedback to that person; and determines the performance rating. The department manager is involved only in determining the dollar amount of the salary increase. Over time, everyone will have an opportunity to serve on the Performance Appraisal Team. In this case, letting go was exercised in the design of the process and structure, which in turn provided latitude in its ongoing implementation. One employee commented, "Everyone in our group feels that the creation of the Performance Evaluation Team was one of our greatest successes for 1994."

Other groups evolved even more direct peer-review processes and came to value them highly after some early growing pains. "We do [peer reviews] very openly now," said one participant, "but it took us a while to get there. It can be hard, but it's a lot better to hear it from someone who knows your work than from a manager who might not." Another said, "Our group is conducting peer-to-peer reviews and they are working. Everybody has taken ownership of the reviewing process."

Involving Teams in Recruitment and Selection

Another human resource system change supporting empowerment evolved with respect to hiring. Team hiring and various processes for involving several levels in giving input to hiring decisions became the norm. This process was cited by an employee as "a good example of empowerment." Each team in her department, she said, "was responsible for screening, interviewing, and selecting candidates. The process was team driven, not management driven. The teams found out that there was a large time commitment involved in hiring, but this increased the ownership of wanting the person to succeed. It also helped build the morale of the team."

Equipping Employees to Exchange Best Practices

The history of Champion's best-practices efforts illustrates the powerful impact information systems can have on style and structure. As part of the Total Quality Management movement, many organizations engaged in "benchmarking" and best-practices studies in the 1990s. Champion's foray into this arena created another opportunity to make the impossible happen. As the executives scanned the mills in 1993, they observed certain mills doing certain things very well and asked why all of them weren't. They established a companywide best-practices team with the executive vice president of publication papers, Joe Donald, as the sponsor. Donald in turn named Terry Womack team leader (interestingly, Womack was the vice president of corporate benefits). The core best-practices team subsequently sponsored subteams in specific areas. The first was the pulp team, followed by a maintenance and engineering team.

They all quickly ran into some powerful organizational norms. Despite progress on so many fronts, competition among the mills still stifled mutual learning and limited total system alignment. People in one mill resented hearing too much about the good things going on in another mill and were not much interested in sharing their own successes either, in part to protect their perceived competitive advantage and in part because they were afraid of seeming boastful. This peer pressure resulted in a cultural norm not unlike the Japanese norm of driving down the nail that sticks its head out or what the Australians refer to as "lopping off the tall poppy." And if it was difficult to share successes, it was almost unheard of to share failures and mistakes. In the past Champion had held "counterpart meetings" for managers from the same function across the mills, but for the reasons just given they had not proved satisfactory. In short, despite so much change in the culture, help was neither as freely sought nor as freely offered across mills as it should have been in a high-performance organization. It was hoped that the best-practices effort would overcome these problems and promote real learning.

According to Womack, the core best-practices team began to work on two issues: "(1) people just plain didn't know each other very well, and (2) they lacked a common vocabulary." To resolve the first problem, the corporate director of organizational development, Bob Knowles, worked with the teams in face-to-face meetings, helping them to get to know one another through such means as telling stories about their mills and looking for the commonalties. In the course of doing so, they discovered the second problem—even though all the pulp-team members made pulp, they articulated their problems and practices differently from mill to mill. In the discussions of commonalties, members frequently had to ask one another, "What do you mean by that?" But when they were asked to go to diagram their process, they discovered they were talking about the same thing in different ways. They then explored the use of comparable measures and ran into cultural norms once again. Publicly shared comparable measures might, after all, lead to comparisons of an odious nature: one's mill might look bad, and "higher-ups" might try to fix the blame rather than encourage problem solving. Measures were finally cautiously agreed on, and performance against those measures cautiously shared. When nothing untoward happened, trust grew and sharing became less guarded. Ultimately, said Womack, the measures became the base of a common vocabulary.

The teams worked diligently but unspectacularly for a while, holding meetings and conferences, enjoying the interchange, and setting some ambitious goals. Yet the cultural barriers did not disappear. The core team searched for a model that would bring rapid learning. In their efforts to understand what was going on—both what was working and what wasn't—they spent some time learning about the diffusion of innovations—for example, the role of change agents and opinion leaders and the horizontal nature of diffusion.[7] Joe Donald arranged a visit for himself and Womack, along with Vice Chairman Ken Nichols and the vice president of management information systems, Dan Smith, to Buckman Laboratories, a supplier of chemicals to the paper industry. Buckman, they found, made extensive use of the CompuServe on-line service to connect their employees worldwide for rapid learning.

Thus Womack was ready when members of the maintenance and engineering best-practices team expressed an interest in finding a better way of staying connected than meetings and conference calls. They discussed an existing e-mail system tied to the Champion voice-mail systems but said that it was too cumbersome, far too time-consuming, and generally not user friendly. Womack promised a solution. The standard procedure would have been to go to the management information systems department with a request. At that time, the response would typically have been to launch a project to develop a new system, which could take months at best. Instead, with Joe Donald running interference, Womack reviewed the best-practices project with the Treetops group and described Buckman's success with CompuServe to show the advantages of taking that route: it was a fairly simple, readily available, and relatively inexpensive solution that could be up and running on a pilot basis in a short time. When the executives asked the cost, Womack was ready with an estimate of about $280,000 for seventy laptop computers, modems, and the CompuServe hook-up. The executives approved it on the spot, by-passing the typical bureaucratic approval steps and with no involvement of management information systems.

About ninety days after the Buckman Labs trip, the maintenance and engineering best-practices team had laptops and the software for connecting with each other on a daily basis. Shortly thereafter the pulp and core teams did as well. Because no one knew exactly what to expect, the team members were told at first just to play with the system. Although goals for the use of the new technology were not made explicit, it was hoped that it would encourage members to share successes and failures, promote rapid learning, encourage creative problem solving, and increase cooperation among the locations, across hierarchical and unit boundaries.

Opinion leaders and corporate executives were encouraged to put messages on the system, with the understanding, Womack said, that "what interests my boss interests me." This informal use helped overcome initial fear, and once the ice was broken, usage began to take off. This was the breakthrough the best-practices effort needed to go from being an interesting activity to being a dynamic exercise in empowerment. It diffused rapidly to a "shrinkage" team that attacked the problems of maximizing yield, to an environmental, health, and safety team, and, getting to the heart of the business, to the papermakers team. This group of opinion leaders (papermakers being the highest-status operating people in any paper company) burned up cyberspace with their queries, shared experiences, and ideas. It helped that they were led by a high-status mill operations manager, Bill Bannan of Courtland, himself an experienced papermaker.

Some of the pull into the process came from the sheer technical novelty of it—receiving a new laptop computer—and from the status—being in the know and its adverse, being afraid of being left out of the loop. More important, early users began to show that real sharing could take place, that learning was enhanced, that problems got solved. Other groups, such as the financial and materials staffs and the Weldwood operations in Canada, got started on CompuServe without the official sponsorship of the core team. The human resources and organizational-development managers went on line in January 1996. In two years the number of connected users went from zero to over seven hundred.

More significant than these numbers, however, is the impact the technology had on furthering the cultural change toward horizontalness and latitude. An engineer in Lufkin, Texas, could now help solve a problem raised on-line by a pulp-mill manager in Quisnel, British Columbia; another idea might come from a manufacturing vice president from another business unit in Stamford, Connecticut, or from a purchasing manager in Mogi Guacu, Brazil.

Members spoke of some early cultural norms they had to overcome. For example, a purchasing manager said she found it easier at first to describe something she or her mill was doing well than to admit a problem. Others experienced just the opposite because they didn't want to appear to be a "know-it-all." Some said the on-line conversations changed when the name of some Stamford executive appeared on the screen, but most said that wasn't the case. Titles and levels didn't matter. Information and solutions were valued wherever they came from; for that matter, members were valued for their willingness to put a problem they were experiencing out to the on-line forum. Even the old definitions of who was an opinion leader began to shift. People who would have held back in face-to-face meetings with some of the "old guard" in the usual positions of influence took the chance of offering ideas on-line, and others responded to them. Other barriers were assaulted. One executive told of seeing a note of congratulations on-line for a promotion that hadn't been officially announced even though it was quite widely known. At first, the executive was angry but then thought to himself, "Hey, that's the way it should work. I can't control it, and I shouldn't try."

Additional challenges included learning the "etiquette" of electronic communications, although some found trivial polite exchanges a waste of their time, and as a result they used the system infrequently. Interaction skills had been learned in the change process for face-to-face meetings; a new level or set of skills had to be developed and learned for this new medium. The cultural norms did not vanish completely or immediately, and there were still perceived political ramifications (Who else is on-line, and what will they—my peers, my boss, my boss's boss—think of my contributions?). Personal concerns had to be dealt with, such as fear of exposure and embarrassment because of lack of education and literacy or ignorance of a particular subject, time required to learn the new skills, as well as time spent on-line instead of doing one's "real job."

Previously perceived barriers—time zones, distance, language, organizational unit, and place in the hierarchy—were reduced however. According to Bob Knowles, a shift was taking place from "I'm tired of hearing about Bucksport [or any one mill]" to "These guys are my allies." New norms of community and accountability for total system success began to evolve. With members from Weldwood of Canada and CPC of Brazil becoming real players in this transnational best-practices exercise, the vision of a truly global company was tangible and real when Champion leaders began to talk of it in the mid-1990s. Knowles observed that "the mental shift toward a total rapid-learning organization, when once embraced, opened up a whole new way of doing things, a whole new way of thinking about things, and a whole new sense of community within the company." The best-practices project both affected and was affected by changes in all of the Three Ss of Letting Go: structure, systems, and style.

As the electronic transfer of knowledge and experience increased, so did the needs of the system; therefore, during 1996 the best-practices forums migrated from CompuServe to more sophisticated Lotus Notes groupware. And, in a strong acknowledgment of the powerful potential of this effort in rapid learning, when Joe Donald took over as the executive vice president for the new manufacturing function in the fall of 1996, he brought Bill Bannan up to Stamford from the Courtland mill in the newly created position of vice president of best practices for all of manufacturing. A year later, in August 1997, Bannan reviewed the goals, achievements, and issues of the best-practices process with senior management; he reported, among other results, a profit impact by the various initiatives across the company of over $12 million per month. Still, he said, learning and transfer of practices was not as fast or effective as it should be, and breaking through cultural barriers relating to autonomy was still difficult. Joe Donald said, "We are still not on top of the rapid transfer and application of knowledge," and he made it one of the top priorities of his manufacturing staff for 1997–1998.

The best-practices effort also affected the management-information-systems department itself. The experience with the best-practices teams, especially the fact that they went around the department and adopted a system already on the shelf, was not lost on the management of the department. This was but one of many changes in both the internal Champion culture and the external information-technology environment that caused the department to take a serious look at itself. After being one of the first groups to go through a redesign in 1988, they undertook another major transformation in their vision and operation in the mid-1990s. They went from essentially a main-frame environment with a focus on the development of systems for the corporate and business-unit executives, who were the major customers, to a "client-server" environment with a vision centered on four elements: partnership, quality solutions, speed, and empowerment. Instead of being expert developers of massive programs, management-information-system teams worked as consultants with their partners in the mills and business units to find the best solutions available and to get them implemented as soon as possible. Through empowering its employees to determine their own resource needs, including their own training, the department could in turn provide products and services that empowered their partners in the field. Within the first year there was a dramatic improvement in the way the department was seen by upper management and managers in the mills. For years there had been complaints that the department was not responsive, had its own "teckie" agendas, and took too long to get a system up and running. A year later their feedback was just the opposite, reflecting a dramatic turnaround in these perceptions.

Within the department employees experienced a dramatic change in latitude. They now were given the authority and the responsibility to manage their own budgets, with, for example, the freedom to order up to $500 worth of software without management approval. Individuals could select the training programs they wanted to attend, inside and outside of Champion, and project teams decided where and when they needed to travel and managed their own travel budgets. One manager who left the company and returned two and a half years later observed, "When I came back, it was a day and night transition. I had a unique opportunity to see it. It has moved from highly controlled to much more freedom." An employee made a similar observation. "In the past fourteen months, I've gone from a manager-controlled environment to a team-controlled one." An employee commented on how it affected project work. "On this project, I was struck by no sense of hierarchy. There was a sense of equality when you walked in the room. This was demonstrated through equal sharing of work load, including grunt work. Our group decided how to distribute the work. We cross-trained each other without specific direction from management. One year ago there was one person who could do support work. Now we all can. The team set its goals and priorities. The facilitator helped on the first day establishing working norms and expectations. Nobody seemed to be working their own agenda. A clear charter at the beginning helped. The team makes decisions about training or equipment, using good common-sense judgment. In the past, management seemed to veto training or equipment requests automatically to save money, without considering the project requirements. The teamwork approach has been easier to work in with less stress. We're starting to work towards peer reviews. The result is that the mills are getting more projects installed faster than they would have [in the past]."

As a result of this transformation a new respect developed for what information systems could contribute to the bottom line. The Treetops group commissioned a high-powered information-systems team in April 1995. They selected members of the team from across the company "to recommend how to use information systems to help Champion . . . run mills better, enhance net revenue, reduce overhead [and] reduce interest costs."

Summary and Conclusions

This sort of pull of a previously impossible vision of the future made to seem possible by the success of the past might be said to characterize Champion as whole as it moved from the mid-1990s into the transition to a new millennium. Champions at every level had demonstrated the power inherent in providing them with new degrees of letting go. Instead of a finite amount of power being driven down, power grew exponentially at all levels, resulting in a much more powerful corporation. The change had not come easily, but ideally it was built to last. In 1996, for the fourth year in row, Champion was selected as one of "America's Most Admired Companies" by *Fortune* magazine and was listed among its fastest gainers from the previous year.

Champion managers—like other managers promoting similar change—had to answer two questions regarding letting go. First, how do we implement letting go? Second, how much do we let go? In answer to the first question, more is better. Just as we observed in the initiatives for achieving alignment and for building capability, redundancy is the key. Champion utilized many redundant signals to empower people, to tell them that they could assume additional power and responsibility.

The second question is more complicated. Letting go sounds simple, but it works only under certain conditions. Individuals or units can be given greater latitude only if they are clear about the goals and other parameters that need to be taken into account (strategic, technological, or social—the alignment touchstones), have developed skills and knowledge, are provided with relevant information, and are committed to the organization and to achieving cooperative relations with each other. Because the level or presence of these preconditions changes, optimum discretion is a moving target. At any point in time, one can aim for the maximum amount of letting go that is justified by the level of alignment and capability. In fact, in order especially to promote further development of capability, supervisors should from time to time provide slightly excess latitude, recognizing they are taking a calculated risk.

The above discussion leads to the conclusion that it is good practice for supervisors and those they supervise to discuss and to make contracts explicitly in the terms used here. How strong is alignment currently? How much capability exists? What are the implications for letting go now and in the future?

Expanding the "What Works" Spiral

As a final look at the interaction of these elements, the expanding spiral in Figure 4.2 illustrates how increases in any of the ingredients of the "what works" formula can support and complement

Figure 4.2. The Expanding Spiral: How the Ingredients Interact.

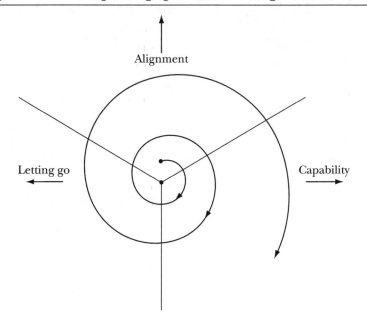

increases in the other two ingredients. The Champion story illustrates how this spiral can expand if the effort is sufficiently sustained and comprehensive. As subsequent initiatives are added to previous ones, they produce increasingly clear, deep, and broad alignment, strengthen capabilities, and allow for effective letting go. Importantly, in Champion there was the tendency for the change effort to move up the hierarchy, contrary to conventional thinking about top-down change. The early focus was on operating employees on the mill floor, the forest floor, and in staff groups. A later cycle involved mill managers and functional heads on multifunctional or multidisciplinary teams and finally included top management and the board of directors. Furthermore, as higher levels of the hierarchy were affected, the process expanded to embrace external constituencies. Whereas the initial focus was on workers and their jobs, it later crossed boundaries within the organization and expanded externally to vendors and customers, and then, finally, to shareholders and the financial community. These trends and sequences over time at Champion are not the only ones that will work, but they are associated at Champion with a remarkably robust and enduring change effort and therefore must be considered instructive.

Epilogue

What would it take to become the best pulp and paper company in the world?
—CEO Andy Sigler, January 1995

Today we begin the task of creating a different Champion.
—CEO Dick Olson, October 1997

When Andy Sigler opened the meeting of the Treetops group in January 1995, it was the beginning of a new kind of dialogue in the company. He reported to the twenty or so senior executives gathered around the highly polished mahogany table that the market had finally turned around in mid-1994 and prices were good and should stay that way for a while; performance of the mills was at record levels; the multi-billion-dollar capital program was at an end; and the balance sheet was in pretty good shape. The two-pronged improvement strategy of modernizing facilities and developing an effective organization had worked fairly well. Champion was competitive in the industry. In a race without a finish line, however, one can celebrate successes but can never declare victory. It was now time, Sigler said, to ask "Where do we go in the future?" "An institution," he declared, "has to find a way to question itself about its core and its direction. We have to create a forum where things are constantly kicked around and the givens are examined." Without meaning to coin a rallying cry, he asked, "What would it take to become the best pulp and paper company in the world?" From that not-so-innocent question grew the theme for an ongoing dialogue lasting over a year.

Without the intentionality of a formal process, "becoming the best" became an ongoing conversation in the company. Although ideas bubbled up and some action plans moved to implementation, there was no attempt to codify a new vision. The dialogue was left open and organic, offering room for both divergence and convergence. This dialogue was still going on when Sigler and Whitey Heist stepped down and Dick Olson, Ken Nichols, and the new executive team took over in August 1996. It became the context in which the new team began its deliberations.

A New Context

Many organizations these days are reexamining their vision, and Champion had done so before. What made this particular dialogue remarkable was that it took place within a profoundly different context than the one approximately ten years earlier that led to the Champion Way statement and the basic change strategy that has been the core story of this book. This new context brought a new spirit to the dialogue, which would not have been possible in an earlier time.

As we discussed in Chapter One, the merger with St. Regis in 1984 resulted in a new context for Champion International: a larger company in the United States with greater economies of scale and with facilities that drove a commodity white-paper strategy, making quality and service essential factors while putting a premium on becoming a low-cost producer. These business and technical factors, combined with the values of the Champion Way, formed the organizational core and became the context for the improvement strategy about which we have written. The process and initiatives that we have covered were aimed at developing alignment with this core and the necessary capabilities and appropriate latitude for Champion to become competitive.

This mid-1980s context energized what Collins and Porras called "trying a lot of stuff"[1] while searching for "what works." We have covered but a small fraction of the multitude of activities aimed at continuous improvement within existing systems as well as others aimed at changing infrastructure, all driven and empowered by the context. The process started at the top of a centralized company and moved downward and outward to the mills and other locations. At some point the locus shifted from high central control to high location control based on differences in local circumstances, culture, and leadership. Successful developments at the local level stimulated change back upward to business-unit and corporate levels. This mutuality of influence took place within a vertical context until, eventually, the horizontal forces began to take hold, and collaborative efforts, characterized by nonhierarchical communications, began to expand across mills and staffs.

We have tried to give a cumulative sense of some of the changes that took place as new webs of inclusion developed, even large gaps and great differences remained from mill to mill, office to office. One answer to the question, then, of what it would take to become the best would be to consolidate, diffuse, and institutionalize those best practices that have been shown to work. This project is certainly important and sufficiently challenging given the nature of any complex organization and human nature itself. Institutionalizing these practices across the board would constitute a major and singular accomplishment in the annals of change management.

Yet the challenge had an even greater scope. As the decade-long process enlarged the vertical and horizontal boundaries in the ways we have described, the entire playing field was shifted outward and upward. Through changing enough of the little pictures, by the mid-1990s Champion found that it had changed the big picture.

Thus, just as a new context was forged in the mid-1980s to drive change, so was that context transformed over a decade, creating a new framework for the new management team as it rigorously attempted to reexamine the alignment touchstones. There are five key parts in the new context.

Intra- to Inter-

First, this transformed context includes moving the focus from intra- to inter-, that is, in looking not only inward but also across. In addition to looking within teams, departments, mills, and business units, the company now looks across the chain of value-adding activities from sources of supply to customers and the customers of customers; to the community and society and the global environment; and to building alliances and partnerships and "virtual organizations" systemically and holistically, embracing these components.

Best on the Planet

Second, from a largely geocentric perspective, the company formerly saw itself as "domestic," with two solid subsidiaries in Canada and Brazil. The "becoming the best" dialogue emanated from a yearning for a decidedly more global perspective that looks outward at markets and competitors anywhere on the planet, with the Canadian and Brazilian operations playing a large part in a synergistic whole.

Low-Cost Producer

Third, the company must be successful financially to be seen as the best; there is deepened recognition that a key to profitability through the ups and downs of a maddeningly cyclical industry is low cost.

Accelerated Change

Fourth, participants in the dialogue, including the new management team, converged around the central point that "our change effort is sound and is a core competency" that provides a competitive advantage; however, they said, this core competency must now be exercised on a larger stage and at an accelerated pace. The foundation built by "making haste slowly" over ten years makes possible (and necessary) greater speed and agility. "What works" is still alignment, capability, and letting go, but on an enlarged scale.

Bold Conviction

Fifth, of all the changes in context, perhaps most striking was the genuine belief that becoming the best was possible, the bold conviction that being the best in the world was in fact within reach. This belief did not exist just a few years earlier. Years of mediocrity had bred an organizational inferiority complex. From 1992 to mid-1994, even as all the internal measures of performance showed a transformed company producing at remarkable levels, the fact that this achievement wasn't reflected in revenues and profits because of depressed industry conditions continued to cause doubt and anxiety. Are we really as good as the numbers show? Will we come out of this stronger than the competition? Or, the refrain went, will we once again snatch defeat from the jaws of victory?

The unmitigated success of the latter half of 1994 and all of 1995 went a long way toward overcoming that fear of success. Participants in the dialogue had become true believers. If you are mediocre, you want to do as well as one or more of your targeted competitors; however, when the context creates an expectation for being the leader of the pack, you can't feed off references to others. You have to set your own standards; you must aspire, as General Electric CEO Jack Welch has been widely quoted as saying, to become "better than the best." There is no one to copy. The ultimate measure of being the best does not exist in the domain of what you do but in the domain of who you are. Champion in 1996 was poised and confident, ready for a genuine, all-out effort to *be* the best.

A New Governing Objective

Poise and confidence are quickly tested in the even larger context of the paper industry. "It's a helluva business," as Nichols said. The white water truly is permanent and unrelenting. By the spring of 1996 the one-and-a-half-year up cycle was over, and pulp and paper prices went into their steepest slide since 1974. What does it mean to be the best in such times? What kind of company will be the best in such a tough business? This was the environment and these were the kinds of questions that had to be addressed by Dick Olson and his Gang of Eight in the meetings you observed as a fly on the wall. The assessment you could make is that the changes at the top had already had effects both inside and outside that group, including

- Elimination of a host of "sacred cows," or business options that in the past had been "off the table"
- Increased openness; indeed, open expression of differences, including sharply felt differences
- Increased ability to think "outside the box"
- Increased focus of middle-level managers on their business segments

- Increased upward influence in the corporate hierarchy
- Increased outward orientation and awareness of customer needs, competitor practices, and shareholder interests

In regard to the final point, Olson would say that summer, after about a year in his new job, "My greatest learnings have come through interfacing with the outside world," a world that included the investment bankers, industry analysts, customers, consultants, and other outside experts who were brought in for presentations to the Gang of Eight. As a result of this learning he and his team determined that an external measure—total shareholder return— would be their "governing objective" for being the best. This decision led his team into the first fundamental analysis of the alignment touchstones since the mid-1980s.

Their deliberations relative to the competitive-strategy touchstone reinforced the belief that it was important to compete on a worldwide stage. Reports from the business teams caused Olson's group to come to grips with the fact that, even with all Champion's new facilities, in many cases they were not the lowest-cost producer in the world because the world had changed and was changing. They had to be concerned with considerable new capacity in Europe, but the real wake-up call was the explosion of new capacity in Southeast Asia. Previously it had been common in Champion and in the U.S. paper industry to ignore this part of the world on the grounds that it didn't have the wood, the skills, or the technology to be a threat. Those grounds had changed dramatically, particularly in Indonesia, where huge new paper machines were coming on line—the world's fastest and most cost effectively constructed machines—and they could be supplied with extremely low-cost wood in comparison with the situation in the United States. It was no small challenge then when Olson declared that "our strategic intent is to be the preferred investment in forest products by manufacturing selected products in the Americas and marketing them to the world."

The business team responsible for core Champion products such as office printing and copy paper (known in the industry as cutsize paper) also determined that the dynamics of that marketplace had markedly changed. Champion had always focused on large customers such as Xerox that were producers themselves rather than end users. Mills and employees were proud of winning Xerox quality awards year after year, feeling secure because they were pleasing their largest customer. However, the era of the personal computer, the local copy shop, and the home office was now here. Superstores such as Office Depot and Staples and warehouse stores were selling cutsize paper directly to small, independent business customers and to the public in general; cutsize had "gone retail." Champion was highly dependent on this product category, which over a few short years had become very volatile. "The whole manner in which uncoated freesheet papers, particularly cutsize, are marketed has changed dramatically," Olson said. "We must reposition ourselves in that business by analyzing and developing very different ways of servicing our customers and perhaps markets that we don't now serve. That business does have excellent growth potential, but we have some significant challenges in front of us to turn it into an economic profit performer for Champion."

Having had their consciousness raised regarding intense global competition and changing dynamics in the marketplace, the team went on to analyze the technology touchstone by reviewing each of their facilities in terms of its potential for creating value, projecting ten years ahead. Creating value was defined as earning an "economic profit"—that is, earning more than the cost of borrowing capital. Champion had been criticized by Wall Street in the past for being too broad, stretched too thin in its product strategy. Internally people thought of that as a strength—"We make everything you can put ink on." As a result of their analysis, Olson said, "we concluded that our product array is too broad. From now on we will focus on the product categories where we believe we have the best opportunity to make economic profit." The technology assessment also showed that competitors had shifted more aggressively than Champion from an acid to an alkaline process for making cutsize and related grades of paper, putting Champion facilities at a cost disadvantage. The team also concluded that, to realize economic profit, the company could no longer invest capital at the rate it had during the past decade.

The character of the conclusions regarding the third touchstone, social values, was quite different from the others in that it emphasized continuity. Olson continually reiterated his ongoing commitment to the Champion Way statement and what he called "a continuing focus on people." As discussed above, the change process was determined to be a "core competency'" and the need was to accelerate it, not cut it back or reverse it. Indeed the new structure of the company was based on the competencies, commitment, and cooperation that had been developed through that process.

In summary, then, the Gang of Eight had looked outside and confronted the realities of an increasingly competitive, dynamically changing, and more global playing field and then, when turning to look inside, had found some serious weaknesses in their market position and in at least some of their facilities. They believed that making changes was a core competency on which to build but that the ability to make rapid change was more important than ever. As September 1997 drew to a close and the deadline for announcements was imminent, Wall Street, customers, competitors, and company employees were awaiting this new team's answers to the questions of what it would take to be the best and what shape Champion would take into the next century.

Building a Different Champion: A Three-Pronged Strategy

The executive team took their plan to the board of directors for a final review in the latter part of September 1997 and made the public announcement of those plans on October 8. In his memo to all employees on that date, Olson began, "In a conference call this morning I advised managers across the company that we have completed our comprehensive analysis of all of Champion's businesses and facilities. As a result we have concluded that we must 1) focus the company on businesses where we can create shareholder value, 2) increase our profitability by improving the profit potential of on-going businesses, and 3) exercise very strong financial discipline in all areas of spending." Each of the three prongs of this strategy is discussed below.

Focus

Dramatic evidence of the commitment to focus on fewer products and businesses came with the announcement of plans to divest several of the company's major mills and other facilities. The newsprint mills in Texas, Sheldon and Lufkin, were put up for sale, as were the specialty mills in Hamilton, Ohio, and Deferiet, New York. The Canton, North Carolina, mill, which made both specialty uncoated papers and bleached board for the liquid packaging business, was also to be sold, along with all the other DairyPak facilities. Finally, most of the timberland holdings in New York, Vermont, and New Hampshire were to be divested as nonstrategic. This left Champion's product focus on coated papers, uncoated freesheet, container board, paper distribution, and softwood timber and lumber—businesses in which it was believed there was the best opportunity to make an economic profit over the cycles of the industry. Olson urged employees to "please keep in mind that we are not closing anything. We believe that these businesses that we are selling are good businesses, operated by good and dedicated people who know their operations well. We are selling them because they do not fit our strategy going forward; and we also believe there are qualified buyers out there who will welcome the opportunity to acquire them." The goal was to sell these facilities by mid-1998.

The six remaining domestic pulp and paper mills were those in Bucksport, Courtland, Pensacola, Quinnesec, Sartell, and Roanoke Rapids. All these mills with the exception of Roanoke Rapids were the same ones listed in Chapter One as constituting a substantial cluster of strategically important mills that had made the formula work, an observation we made before knowing the outcomes of the strategic analysis by the executive team. In all these mills there had been extensive capital investment in the operations, and all had implemented extensive systemic change over the years to support a high-commitment culture through alignment, capability, and appropriate letting go.[2]

This is not to imply that these were the reasons they were retained; on the contrary, the criteria for those decisions clearly had to do with the potential for an economic profit in the product lines and businesses encompassed by those facilities. Nevertheless, it is possible to speculate that their aggressive and comprehensive embracing of the change process may have contributed to that potential, and, more vitally for the future, it is encouraging for the company to start this next phase of its life with facilities and organizational cultures that are far ahead of the norm.

Profit Improvement

Having determined the configuration of facilities and businesses that they would take into the future, the executive team outlined strategies for improving profit in those ongoing businesses; these strategies included growth by acquisition or expansion, leveraging certain assets, repositioning some product categories in the marketplace, and improving customer focus. They set a target of $400 million annual savings by the year 2000 through this repositioning along with productivity increases and further reductions of about two thousand jobs worldwide in the remaining locations.

"These reductions," Olson told employees, "will be handled in a way that is consistent with the Champion Way and in a fair and equitable manner. Our pledge to you is that we will strive to bring Champion and its people through a difficult time with the utmost understanding, compassion, and fairness." Mark Childers, senior vice president of organizational development and human resources, also emphasized that the Champion Way statement would be the guide for a "collaborative process" for handling the reductions.

Financial Discipline

Whatever good things Champion had become known for over the previous decade, financial discipline was not one of them. As we have covered previously, billions of dollars were spent in that time to improve facilities, spending that Wall Street often criticized. Now the company was faced with the daunting task of earning an economic profit on those huge investments. The new commitment was to bring future capital spending down to the level of depreciation. No new pulp mills or paper machines would be built in North America in the foreseeable future, although it was possible that such expansion might take place in Brazil. All expansion possibilities, including potential acquisitions, would be undertaken only when there was a real opportunity to "create value"—that is, earn an economic profit—with careful attention to not overpaying. Alliances and joint ventures would be pursued as a way to improve profitability while minimizing capital investments. As Manufacturing Executive Vice President Joe Donald put it, "We need to move from a capital-intensive business to a capital-intelligent business."

In addition to the close watch on capital spending, Donald emphasized the need for financial discipline in operations. He said that the work of the business teams had forced the company to come to grips with the fact that the world had changed and that Champion, even with all its new and rebuilt machines, was not the low-cost producer, especially given the expansion in Southeast Asia. The company needed to move aggressively to alkaline rather that acid production processes and more generally to become "obsessed with reducing costs, shifting to a low-cost mentality." He also put the need to become a rapid-learning organization high on his list of priorities for becoming cost effective.

Initial Reactions

With this new strategy emphasizing improving total shareholder return as the governing objective, it was natural to turn to Wall Street for initial reactions to the announcement. Olson, Nichols, and others had been meeting with analysts and other representatives from the investment community right along, and the thinking and principles behind the new strategy were not surprises; expectations had already been incorporated into the price per share, which had been rising over the year. Nevertheless, the response to the announced plans was generally positive. Typical comments were "Champion takes the long view" with its "clear and compelling strategic direction"; "most aggressive and strategically focused"; "more focused . . . and more agile"; "hits the mark." About the only reservation was that some of the changes, though on target, were overdue. Customers also generally liked what they saw and heard. The share price moved up over the next few months.

The communities with Champion mills on the sale block were not so sanguine. They quite naturally reacted with concern when the announcements were first made, anticipating the worst. However, when it became clear that no plant closings were foreseen and that there was some optimism that appropriate buyers could be found, they quickly rebounded and went to work with the company to enhance the attractiveness to potential purchasers.

Likewise employees and unions had some fears. In the facilities to be divested, there were mixed feelings. In some mills people had expected the news for some time and felt relieved when it was finally official. For many it was not the first time they had been through a sale, and they were confident they could survive this one, although there was always concern that some unknown new owners might make further employment cuts. Some of those mills took it as a challenge to show potential new owners what a good organization they were. Employees in some mills came to that stance only after their initial anger at Champion for "abandoning" them abated. There was talk in some locations of organizing an employee purchase of the mills; in most cases that quickly died down, but in at least one instance it was still being actively explored as of this writing.

In the retained mills the mixed feelings included relief at not being sold yet worry about the pending job cuts. Shortly after the first public announcements, Joe Donald and Mark Childers flew to the Nashville headquarters of the UPIU to meet with President Boyd Young and some of his officers. In previous meetings over the past summer, Champion had briefed union leadership on the basic issues involved and the general direction to be taken, even though the specifics had not yet been decided. Therefore the union was not caught completely by surprise by the announcements even if it was deeply concerned with the total scope of the final restructuring. While expressing his concerns, Young assured the company that the union understood the industry and the issues it was facing and was committed to working with Champion on these issues in the future.

The union and the company went ahead with the December Forum meeting (as discussed in Chapter Three), which was aimed at taking "the first steps in planning the next generation of the Champion-UPIU partnership for mutual success." Significantly, however, only representatives of the retained mills participated since they were the mills that would be part of this next generation. The meeting was held over one night and the following day in the UPIU's Nashville headquarters building.

The first evening Olson gave an overview of the business environment and the three-pronged strategy for "maximizing total shareholder return," emphasizing the need for help from the Unions and all employees "to become the kind of company we have to be." Young then addressed the group thoughtfully about the need for change if the paper industry, which he considered to be in crisis, was to survive. He said he was sometimes asked whether change was as essential as everyone seemed to be saying it was and his answer was no; "change is not essential simply because survival is not mandatory." His second and more serious answer was that the paper market is truly a global one and is becoming more competitive all the time. "Unless we change, we will be left behind like the steel industry, the garment industry, and the electronics industry." He said that many changes would be required including a "cultural change between labor and management wherein true partnership [is] developed."

In response to questions about "What's in it for labor?" Young expressed his hope that a partnership with management in the change process would result in "a stronger, more capable union; secure jobs, wages, and benefits for union membership; a constructive workforce; respectful and dignified treatment; an opportunity for my membership to share the gains created by the new concept; recognition that a unionized workforce properly motivated and empowered adds value and can outperform any workforce in the world."

That first evening was capped with the ceremonial signing by Olson and Young of a revised Joint Statement of Principles, which carried over most the language of the statement quoted in Chapter Three while strengthening the second principle, which has to do with the company's acceptance of "the legitimacy and the institutional integrity" of the union. New language was added to clarify that principle as it applied to how the parties would handle attempts by the UPIU to organize employees in Champion's few nonunion locations; the new language emphasized mutual respect and the avoidance of negative campaigning by either party. The public signing of this document by the two leaders symbolized a renewal of the solid relationship developed in the recent past as the partners moved toward creating a new and uncertain future together.

The group met for most of the next day. Joe Donald covered in some detail the external and internal analysis the Gang of Eight had gone through in arriving at its new strategic direction, and Mark Childers talked about the process the company was going through for restructuring the salaried workforce. He invited union input into a similar process for the represented employees, which would begin in early 1998, and he emphasized the need for open communications and the desire to work through the process in a cooperative spirit with the unions. In his remarks, Vice President Don Langham of the UPIU, who had spent considerable time in union activities overseas in recent years, underlined the reality of international competition. The meeting participants discussed the need to continually clarify and develop the concept of partnership and selected a task force to recommend a model of what the next stage of partnership should look like. They then set dates for the next Forum meeting and expanded the membership to permanently include mill operations managers and international union representatives. Thus, in the context of difficult times that would test any union-management relationship, the partners committed to move on together, recognizing that the difficult times were just beginning.

Moving on in Difficult Times: Some Open Questions

Unfortunately Champion's announced restructuring was not an isolated case in the recent history of American business and industry, and severe downsizing was not limited to the "chain saws" of the business world. Companies generally considered enlightened and humane have painfully had to face up to new global realities. Not too many years ago, IBM began a major reversal of its decades-old no-layoff policy, and Xerox cut ten thousand jobs for the first but not the last time. In the autumn of 1997, along with Champion's announcement, Kodak declared massive reductions and was criticized by Wall Street for not cutting deeply enough. Kodak CEO George Fischer, who apparently had been trying to avoid any cuts, said his critics were being irresponsible. But what if taking the tougher stance called for by Wall Street might in the final analysis save Kodak as a company and preserve more jobs than Fischer's more humane values? Indeed, Kodak later announced further reductions. Levi Strauss, like Champion, was also engaged in a major organizational-redesign and cultural-change effort aimed at employee empowerment. It joined the tough-times chorus with Champion and Kodak in the fall of 1997 with the announcement that it was closing eleven domestic plants because of competitive market conditions. All these decisions were made at the same time that the U.S. economy was enjoying long-term steady growth, with unemployment near record lows, productivity near record highs, and virtually no inflation—all of which bolstered a long-term bull market on the U.S. stock exchanges. It was a difficult and confusing time to be making decisions about the future of a company (or, for that matter, a union).

As they moved on to implementing their new strategy in turbulent times, Champion leaders faced many dilemmas, and many questions as to how to proceed were still unanswered; these questions included but were not limited to the following:

- Will management, even with input from employees and unions, be able to do this cutting and still maintain and continue the culture that was built so painstakingly over the previous decade?
- Will the new emphasis on "total system optimization" across the mills and the call for a "total systems mind-set" mean moving decisions up the ladder toward more centralized control and decision making when, for the last ten years, the struggle was to move much decision making down and to encourage location and mill-floor autonomy? In other words, will the new emphasis change the parameters for letting go? Will it affect employee commitment and motivation?
- Much had been learned in the change process over the previous decade by allowing locations the freedom to experiment and by *not* standardizing across the mills. Is the culture now in a new phase of the learning curve where standardizing will further rapid learning, or, indeed, will it inhibit such learning?
- Will the focus on total shareholder return as the governing objective change the way the company relates to other stakeholders such as employees, unions, and communities? Can this objective become the basis for the creation of the sort of positive imagery that aligns an organization and its people and pulls them rapidly and effectively toward higher possibilities? How does the constraint of this new focus create tension between strategy and social values, and how will that tension be resolved?

The ingredients in the framework we have used for this book should still hold validity for dealing successfully with these dilemmas. Surely, if it is to be the best, Champion should continue to strive to become a company in which alignment exists among its competitive strategy, its technology, and its values, as well as a company in which employees at all levels are aligned with being the best, have the capabilities to perform at the highest levels, and are given the latitude to apply their skills and dedication to the quest. The ultimate open question is this: Will there continue to exist the "indefatigable will" to sustain a change process as comprehensive in scope and as persistent and unrelenting as the challenges to be faced?

Appendix A: Research Methodology

The three of us decided in 1993 to write a book documenting and analyzing the change process at Champion. Prior to that decision, our own observations of Champion during the first seven years of the journey reported here were those of participants—participants who cared deeply whether the change effort would gain momentum rather than lose steam, whether it would extend to parts of the corporation not yet deeply involved, and whether it would make a big difference to the competitiveness of the enterprise and to the well-being of its members. We were anything but detached observers.

After undertaking the book project, we continued of course to care deeply about the robustness of the change process (and two of us especially, Ault and Childers, continued to contribute to it in various ways), but we also began to work on another track. We attempted also to observe, assess, and critique the change effort at Champion the way detached researchers of any other organization would.

In effect, we sought the best of the worlds of both practitioners and researchers. We endeavored to exploit all the advantages of "having been there" as participants while achieving as much as possible the objectivity of detached observers. In this effort, Ault and Childers, as consultant and manager, contributed the lion's share of observations of change events in the organization. Walton, who had a brief role as consultant to Andy Sigler and his staff at the beginning of the period, subsequently observed the changes from a distance as a member of the board of directors. Given his relative detachment and his training and experience as a social scientist, Walton accepted a greater responsibility for ensuring that enthusiasm for the change effort didn't swamp our collective ability to describe it and assess it with some degree of objectivity.

Below we list several aspects of the situation and of our data-gathering and sense-making activities, each of which enabled us in our attempts to exploit our close proximity to the subject and to achieve some degree of detachment from it.

- Our documentation of the change (or lack thereof) in every mill and other parts of the organization was extensive, producing thick descriptions of the changes and assessments of their effects. These reports, many of them prepared by Ault, were reviewed and critiqued for their accuracy as well as their action implications by members of the subject organizational unit.
- We relied heavily on hard-number indicators to corroborate our subjective assessments of the change in soft variables like attitudes and relationships.
- We asked for and received from many managers critiques of the book manuscript to be sure our facts or interpretations jibed with their understanding.
- Because the overall story is one of successes, we have consciously endeavored to identify the tactical failures wherever possible and give them as much attention as their importance deserved.

Appendix B: The Champion Way Statements

The Champion Way

Champion's objective is leadership in American industry. Profitable growth is fundamental to the achievement of that goal and will benefit all to whom we are responsible: shareholders, customers, employees, communities, and society at large.

Champion's way of achieving profitable growth requires the active participation of all employees in increasing productivity, reducing costs, improving quality, and strengthening customer service.

Champion wants to be known as an excellent place to work. This means jobs in facilities that are clean and safe, where a spirit of cooperation and mutual respect prevails, where all feel free to make suggestions, and where all can take pride in working for Champion.

Champion wants to be known for its fair and thoughtful treatment of employees. We are committed to providing equality of opportunity for all people, regardless of race, national origin, sex, age, religion, disability, or veteran status. We actively seek a talented, diverse, enthusiastic workforce. We believe in the individual worth of each employee and seek to foster opportunities for personal development.

Champion wants to be known for its interest in and support of the communities in which employees live and work. We encourage all employees to take an active part in the affairs of their communities, and we will support their volunteer efforts.

Champion wants to be known as a public-spirited corporation, mindful of its need to assist nonprofit educational, civic, cultural, and social welfare organizations which contribute uniquely to our national life.

Champion wants to be known as an open, truthful company. We are committed to the highest standards of business conduct in our relationships with customers, suppliers, employees, communities, and shareholders. In all our pursuits we are unequivocal in our support of the laws of the land, and acts of questionable legality will not be tolerated.

Champion wants to be known as a company which strives to conserve resources, to reduce waste, and to use and dispose of materials with scrupulous regard for safety and health. We take particular pride in this company's record of compliance with the spirit as well as the letter of all environmental regulations.

Champion believes that only through the individual actions of all employees—guided by a company-wide commitment to excellence—will our long-term economic success and leadership position be ensured.

The Champion Way in Action

To help us meet the challenges of the future we have developed "The Champion Way in Action."

It reaffirms the enduring values of the Champion Way philosophy and explains how we will apply them more effectively in our daily operations by changing how we organize and manage work.

Why We Must Change

Champion has the potential to be a great company. However, if the company is to survive—let alone hold a leadership position—our profitability must be enhanced.

Our goal is to achieve and maintain a level of profitability that places us in the upper quartile of U.S. industry, a position achieved by top performers in our industry. Reaching this ambitious goal will enable us to

- Accelerate the modernization of our operations
- Serve customers more effectively
- Improve opportunities for developing the full potential of employees
- Increase return to shareholders
- Enhance continued employment opportunities

Without dramatic improvement in our profitability, there will be neither continued growth nor job security for any of us.

We operate in a fiercely competitive world marketplace against determined and skilled competitors. To outperform the competition, we must

- Increase productivity
- Reduce costs
- Improve quality
- Strengthen customer service

Changing the Way We Work

The challenge we face dictates the need for radical change in our traditional ways of doing things.

It requires a new dimension of intelligent participation from all Champions. Ours is a capital-intensive industry in which the competence and dedication of people make the difference. All employees need to become better problem solvers, learn to work more closely with others, and have the information to make informed decisions. The company, in turn, will support these efforts by spending capital dollars for profit-improvement projects which make strong economic sense.

The purpose of "The Champion Way In Action" is to guide how all Champions work together so that our combined skills, knowledge, and experience are utilized fully.

The statement is based on the conviction that Champions have more to contribute—collectively and individually—than our traditional management practices have allowed. When properly motivated and given enough flexibility in their jobs, people who participate in making and carrying out decisions become more effective because they

- Contribute more ideas
- Accept more responsibility
- Are more satisfied with their work

To stimulate this kind of participation requires an atmosphere of trust.

We must establish a new working environment which eliminates bureaucratic obstacles. We must promote shared understanding and objectives and directly involve employees in operating decisions. Clearly these changes won't occur overnight. They will take time, patience, and commitment.

By changing the way we organize work and the way we manage, we believe that we can gain the needed edge in quality, service, and productivity while increasing job satisfaction.

The potential value—both in profitability and in human fulfillment—of more effective individual and group performance is enormous.

Our Guiding Management Practices

To create the working environment we seek, we will be guided by the following management practices:

- Communicate fully so that everyone understands the business environment in which we are competing.

  The company will share information on performance against standards, financial results, business conditions, company goals, and plans for the future. At the same time, each of us must become a better listener.

- Work to establish and preserve an atmosphere of trust in which everyone can be heard, differences can be aired, questions asked, and conflicts resolved.

  Everyone is expected to help create working conditions which foster cooperation, teamwork, openness, flexibility, innovation, and individual initiative.

- Train and retrain employees at every location to be certain that everybody is qualified to fulfill his or her complete responsibilities and to help everyone learn new ways of working together.

  Changing the way we work will succeed only if people have the understanding, skills, and desire to make it happen.

- Hold leaders at every level responsible for the development of those who report to them.

  Through appropriate delegation, they must build the decision-making skills and self-esteem of those who report to them without abdicating their own responsibility to provide decisive leadership.

- Expect from all Champions a willingness to accept more responsibility.

  All of us have an obligation to do our best to help the company succeed. Working with one another in a new spirit of participation means accepting more responsibility and will enable the company to meet measurable performance goals.

- Assure all employees that no one will lose employment because of work redesign.

  Reductions in force which result directly from implementation of employee work redesign projects will be handled through attrition, voluntary severance programs, or reassignment.

- Explain to all employees that in all other situations where jobs must be eliminated, the company will make a good-faith effort to do what is right for the business and what is fair for the employees.

When organizations, capital improvements, a change in market conditions, or other circumstances cause a temporary or permanent decrease in jobs, the company will try to handle those reductions by attrition, voluntary severance programs, or reassignment. However, there may be conditions under which layoffs or terminations cannot be avoided.

- Establish an atmosphere in which promising new ideas are encouraged and can be tested.

It is a responsibility of leadership to encourage risk and experimentation that may produce benefits to the company even though some promising initiatives will fail.

Summary

Champion's goal of profit performance in the top quarter of American industry is realistic and attainable.

Achieving that goal is everyone's best assurance of continued employment and the chance to share in the results of our efforts. We believe that we can meet the challenge by changing the way we work to encourage maximum participation by employees.

Appendix C: Gainsharing

Figure C.1. Readiness Assessment.

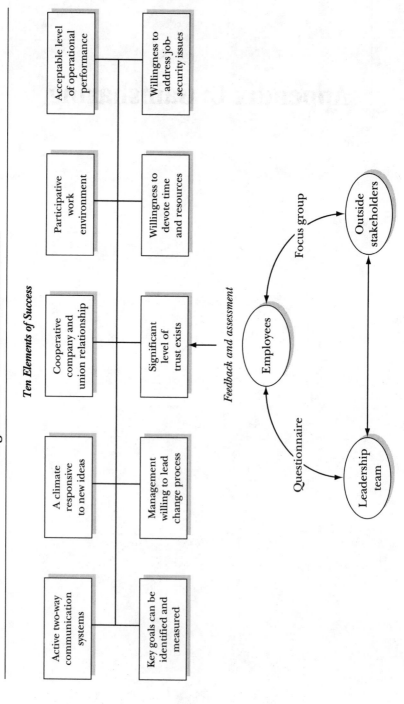

Ten Elements of Success

Figure C.1. Readiness Assessment (continued).

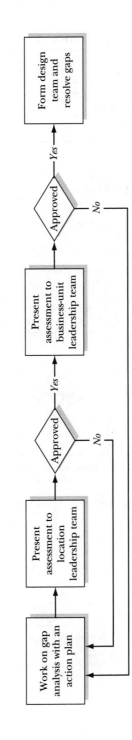

Figure C.2. Process.

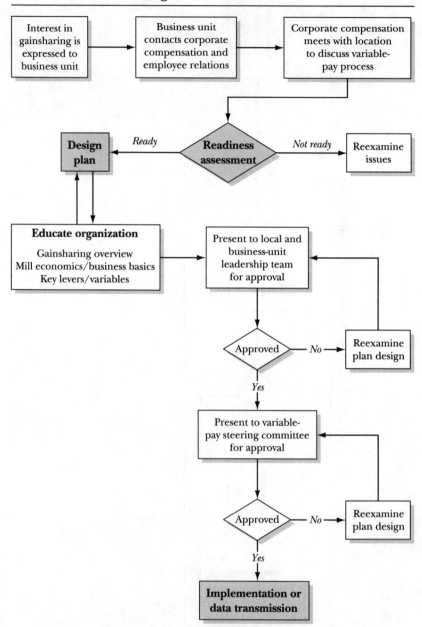

Figure C.3. Design Plan.

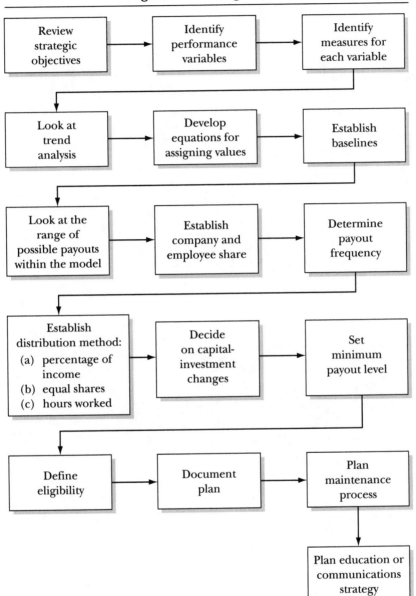

Notes

Preface

1. J. C. Collins and J. I. Porras, *Built to Last: Successful Habits of Visionary Companies* (New York: HarperCollins, 1994).
2. G. Closset and M. Feigenbaum, "Designing a Technical Organization for High Performance," *Tappi Journal,* Feb. 1992, *76*(2), pp. 73–80.
3. R. E. Walton, J. E. Cutcher-Gershenfeld, and R. B. McKersie, *Strategic Negotiations: A Theory of Change in Labor Management Relations* (Boston: Harvard Business School Press, 1994), pp. 67–77; J. E. Cutcher-Gershenfeld, R. B. McKersie, and R. E. Walton, *Pathway to Change: Case Studies of Strategic Negotiations* (Kalamazoo, Mich.: W. E. Upjohn Institute for Employment Research, 1995), pp. 143–169.
4. J. R. Katzenbach, *Teams at the Top* (Boston: Harvard Business School Press, 1997), p. 39.
5. A. Dunlop with B. Andelman, *Mean Business: How I Save Bad Companies and Make Good Companies Great* (New York: Random House, 1996).
6. N. Tichy, *Transformation Leadership* (New York: John Wiley and Sons, 1986).
7. R. H. Miles, *Corporate Comeback* (San Francisco: Jossey-Bass, 1997).
8. R. H. Miles, *Leading Corporate Transformation* (San Francisco: Jossey-Bass, 1997).
9. D. Nadler and D. Kearn, *Profits in the Dark; How Xerox Reinvented Itself and Beat Back the Japanese* (New York: HarperCollins, 1992).

Prologue

1. Katzenbach, *Teams at the Top,* p. 39.

Chapter One

1. Quoted in T. BeVier,"UP Plant Pulps Old Management Theories," *Detroit Free Press,* 1986, pp. 1, 7.

2. M. Beer, R. A. Eisenstat, and B. Spector, "Why Change Programs Don't Produce Change," *Harvard Business Review,* Nov.-Dec. 1990, p. 166.

Chapter Two

1. Closset and Feigenbaum, "Designing a Technical Organization for High Performance," p. 79.
2. Closset and Feigenbaum, p. 79.
3. Beer, Eisenstat, and Spector, "Why Change Programs Don't Produce Change," *Harvard Business Review,* Nov.-Dec. 1990, p. 166.
4. H. C. Metcalf and L. Urwick (eds.), *Dynamic Administration: The Collected Papers of Mary Parker Follet* (London: Pitman, 1941); M. Wheatly, *Leadership and the New Science* (San Francisco: Berrett-Koehler, 1993).
5. R. W. Jacobs, *Real Time Strategic Change* (San Francisco: Berrett-Koehler, 1994).
6. Jacobs.

Chapter Three

1. Macy, telephone conversation, November 1994.
2. F. Herzberg, B. Mausner, B. Snyderman, *The Motivation to Work* (New York: John Wiley and Sons, 1959).
3. Walton, Cutcher-Gershenfeld, and McKersie, *Strategic Negotiations,* p. 80.
4. Walton, Cutcher-Gershenfeld, and McKersie, p. 88.
5. Walton, Cutcher-Gershenfeld, and McKersie, pp. 350–351.

Chapter Four

1. F. Emery and M. Emery. Adapted by R. Rehm from *Participative Design for Participative Democracy* (Canberra: Centre for Continuing Education, Australian National University, 1993), p. 2.
2. Beer, Eisenstat, and Spector, "Why Change Programs Don't Produce Change," *Harvard Business Review,* Nov.-Dec. 1990, p. 159.
3. Beer, Eisenstat, and Spector, p. 159.
4. F. Emery, "Participative Design: Effective, Flexible and Successful, Now!" *Journal for Quality and Participation,* Jan.-Feb. 1995, p. 6.
5. Emery, p. 6.
6. Emery, p. 6.
7. E. M. Rogers, *Diffusion of Innovations* (2nd ed.) (New York: Free Press, 1983).

Epilogue

1. Collins and Porras, *Built to Last,* p. 140.

2. In that same discussion we described Roanoke Rapids as strategically isolated from the mainstream of Champion, and, indeed, this was one of the few aspects of the newly announced plan that was questioned by Wall Street, which wanted to know how Roanoke Rapids fit into the company's focus. However, we did, in Chapter One, acknowledge this mill's solid alignment of its total organization to its customers' needs, cooperative labor relations, tradition of competence, and good work ethic; these factors contributed to its consistent profitability and outstanding safety record.

Index